SpringerBriefs in Complexity

Springer Complexity

Springer Complexity is an interdisciplinary program publishing the best research and academic-level teaching on both fundamental and applied aspects of complex systems—cutting across all traditional disciplines of the natural and life sciences, engineering, economics, medicine, neuroscience, social and computer science.

Complex Systems are systems that comprise many interacting parts with the ability to generate a new quality of macroscopic collective behavior the manifestations of which are the spontaneous formation of distinctive temporal, spatial or functional structures. Models of such systems can be successfully mapped onto quite diverse "real-life" situations like the climate, the coherent emission of light from lasers, chemical reaction-diffusion systems, biological cellular networks, the dynamics of stock markets and of the internet, earthquake statistics and prediction, freeway traffic, the human brain, or the formation of opinions in social systems, to name just some of the popular applications.

Although their scope and methodologies overlap somewhat, one can distinguish the following main concepts and tools: self-organization, nonlinear dynamics, synergetics, turbulence, dynamical systems, catastrophes, instabilities, stochastic processes, chaos, graphs and networks, cellular automata, adaptive systems, genetic algorithms and computational intelligence.

The three major book publication platforms of the Springer Complexity program are the monograph series "Understanding Complex Systems" focusing on the various applications of complexity, the "Springer Series in Synergetics", which is devoted to the quantitative theoretical and methodological foundations, and the "SpringerBriefs in Complexity" which are concise and topical working reports, case-studies, surveys, essays and lecture notes of relevance to the field. In addition to the books in these two core series, the program also incorporates individual titles ranging from textbooks to major reference works.

More information about this series at http://www.springer.com/series/8907

Hassan Qudrat-Ullah

The Physics of Stocks and Flows of Energy Systems

Applications in Energy Policy

 Springer

Hassan Qudrat-Ullah
York University
Toronto, ON
Canada

ISSN 2191-5326   ISSN 2191-5334   (electronic)
SpringerBriefs in Complexity
ISBN 978-3-319-24827-1   ISBN 978-3-319-24829-5   (eBook)
DOI 10.1007/978-3-319-24829-5

Library of Congress Control Number: 2015950885

Springer Cham Heidelberg New York Dordrecht London

Printed on acid-free paper

Springer International Publishing AG Switzerland is part of Springer Science+Business Media
(www.springer.com)

*To my wife, Tahira Qudrat, who patiently
listens to my ramblings about complex
systems and bears with the associated,
often logical but long discussions,
and my wonderful children, Anam, Ali,
Umer, and Umael, who, on an ongoing
basis, represent a beautiful and adorable
kind of a complex dynamic system.*

Preface

People from all walks of life regardless of their origin, creed, sect, social strata, and value systems desire a better life. Energy plays a fundamental role in the well-being, prosperity, and improved lives of all of us. Planners and decision makers of any region, state, or country therefore, have or at least aspire to have a common goal: an adequate, reliable, cleaner, and affordable supply of energy. In order to achieve this goal, the responsible governments and authorities, often at the time of the promulgation of their national development plans and economic agendas, enact, implement, assess, and adjust various energy policies.

To realize energy policy objectives, the common challenge for the decision makers is how to design an effective and efficient energy policy subject to sociotechnical constraints. To begin with, the design of an effective and efficient energy policy is a complex task. Energy systems are essentially sociotechnical systems. Although the complexities of physical energy generation systems (e.g., run-of-river and reservoir-based hydro power plants, nuclear power plants, wind, solar, biomass, and tidal power generation technologies, transmission and distribution systems) are highly technical, the nature of issues related to the affordability (e.g., ability of the consumers to pay) and adoption of technologies (e.g., due to the consumers' sensitivity to environmental emissions) is predominantly social. The limited resources and conflicting objectives of various stakeholders (e.g., local community, business community, regulatory agencies, and governments) further complicate the energy policy decisions. Therefore, an appreciation and understanding of the dynamics of these sociotechnical systems is an essential prerequisite of the design of an energy policy. Hence, the objective of this book is to enhance systematically our understanding of and gain insights into the general process by which the physics of stocks and flows of an energy system explains the making of an efficient and effective energy policy.

Approach of This Book

To realize the objective of this book, we identify some fundamental accumulation processes (stocks and flows) spanning across the demand and supply sectors of energy systems that are common to the majority of energy policy related issues across the globe. To understand the physics of these identified stocks and flows of energy systems, we apply the system dynamics approach (i.e., stocks and flows based modeling). By presenting examples of some "validated and in use" system dynamics simulation models wherein these accumulation processes are drivers of the behavior of the represented energy system, we demonstrate the importance and utility of the fundamental and foundational structures (i.e., stocks and flows based feedback loops) of energy systems. Based on these modeling efforts, three cases, (i) the dynamics of green power in Ontario, Canada, (ii) socioeconomic and environmental implications of the energy policy of Pakistan, and (iii) the dynamic of the electricity generation capacity of Canada, are analyzed and discussed. Finally, for dealing with complex, dynamic energy policy issues, a step-by-step model based on the stock and flow perspective is presented.

Outline of Book

This book has seven chapters dealing with three distinct but related themes of energy policy:

1. The introduction to the subject of "modeling for energy policy," its importance and complexity, and stock and flow perspective (i.e., system dynamics approach) as a "solution" modeling approach (Chaps. 1 and 2).
2. The physics of stocks and flows and how it deals with the complexity of energy systems (Chaps. 3 through 5).
3. The physics of stocks and flows in action; the applications of system dynamics models developed for dealing with various energy policy issues, thereby, helps to drive critical lessons to be learned (Chaps. 6 and 7).

Use and Users of Book

This book provides the reader with a comprehensive understanding of the physics of stocks and flows encountered in the design of an efficient and effective energy policy for a region or country. By studying the dynamics of the fundamental structures of an energy system, she or he will not only appreciate the existence and utility of key stocks and flows of energy systems but also contribute to the policy-making projects across various domains. This book is intended for managers and practitioners, teachers, researchers, and students of design and assessment of energy policy making for complex, dynamic energy systems.

For managers (utility managers, professionals working in energy, environment, and resource industries) and practitioners (energy policy consultants, climate change writers and journalists, and IPPs consortiums), this book provides insights into the complex dynamic systems that they often encounter. For these readers, Chaps. 4 and 5, which provide methodological details on the physics of stocks and flows, may be skipped initially so that they can go directly to the applications of stocks and flows based simulation models to various energy policy related issues and the important lessons learned (Chaps. 6 and 7).

Teachers and instructors in the post-secondary education sector can adopt this book as a textbook for a half course on Policy Modeling or Energy Policy Modeling.

For researchers and students, this book provides probably the most systematic study of the physics of stocks and flows in energy systems. The background material in Chaps. 3 through 5 provides a solid base to understand and organize the existing system dynamics models for energy policy design.

Toronto, Canada   Hassan Qudrat-Ullah
August 2015

Acknowledgments

First of all, I thank Almighty Allah (SUT) for granting me the faculties to embark on this journey. For any project you undertake, you need a constant source of inspiration, encouragement, and motivation. I am fortunate to have my father, Safdar Khan, the very source of all of it: at 94+ years of age, his concerns and sacrifices for the well-being of "others" provides me with huge energy and shines my way. My mother, Fazeelat Begum, travels across the Atlantic to provide selfless and special moral support.

Contentwise, this brief book draws heavily on my own research in and practice of "energy policy modeling" that I started in 1996–1997 at the University of Bergen, Norway, during my masters' program; I am grateful to Prof. Paal Davidsen for introducing me to this research area, specifically for teaching system dynamics methodology.

I would like to thank the colleagues from Springer: Christopher, and HoYing, for their encouragement and support throughout this process. Ms. Vani of Scientific Publishing Services (P) Ltd. is appreciated for her professional support and timely production of the camera-ready copy for this book. I am thankful to Anam Qudrat and Ali Qudrat, who provided the professional editorial support for this manuscript.

The other people, who I specially want to thank for their prayers and good wishes for me throughout my academic and professional life include: my sister, Zahida, and brothers, Naheem, Saleem, Naveed, and Wasim. Tahir's support in my several mundane tasks is also appreciated. Last but unparalleled prayers and support of Saira Bano and Allah Ditta are worthy of mention here. Finally, Col. Tariq (late) will always be remembered for his support of me, especially during the very challenging time of my early (student) life.

Contents

Chapter 1
Energy Policy Making: A Complex Dynamic Task

> We can no longer continue with a status quo energy policy. We must create sustainable clean energy jobs and leave the planet to our children and grandchildren in better shape than we found it
> —Jeff Merkley

1.1 Introduction

In our modern-day life, energy policy has always been at the forefront of public debates, legislative fora, and public and corporate decision making. Decision makers constantly have to find ways to balance the two critical aspects of any energy policy: economic welfare and environmental degradation. This is not an easy task. There is a basic dilemma in energy use. Energy use can contribute to economic growth and development activities, while at the same time, it is one of the biggest sources of environmental emissions that pose a serious threat to the sustainability of our planet. Researchers, especially the modeling and simulation community, continue to investigate and provide policy insights on dealing with these complex issues.

For me, there are three fundamental concerns that have provided motivation to bring out words in this endeavor. First, the need for energy for improving (e.g., in the case of the majority of developing countries) and maintaining (e.g., in the case of industrialized and developed nations) our livelihood and well-being has reached the level of unprecedented dependency. Second, the security of the energy supply in our globalized world is facing unsurmountable risks. Third, the effects of climate change (e.g., increased CO_2 emissions) due to the burning of fossil fuels have created enormous challenges for all nations. In the context of these concerns, the design and assessment of energy policies is not only a worthwhile research activity but can also play a major role in shaping the way we produce and consume energy.

© The Author(s) 2016
H. Qudrat-Ullah, *The Physics of Stocks and Flows of Energy Systems*,
SpringerBriefs in Complexity, DOI 10.1007/978-3-319-24829-5_1

Energy systems are stocks and flows of limited resources including capital (i.e., physical production machines and input materials), money (i.e., payment and receipts for goods and services), information (i.e., management decisions, processes, and reports), and people (e.g., the stakeholders) that are causally interconnected. The interactions of these stocks and flows create the resulting outputs (e.g., the gaps in supply and demand of electricity, and changes in electricity prices). When it comes to understanding of the dynamics of stocks and flows of a system, the empirical evidence is mixed at best (Sterman and Sweeney 2002; Pielke et al. 2008; Cronin et al. 2009). Although the nature of energy policy related issues is complex, the essence of several complex problems such as global warming is as simple as filling a bathtub. The stock of water in a tub is increased by the inflow (i.e., opening the faucet) and is decreased by outflow (i.e., opening the drain). Now if people are given information on both the inflow and outflow, it is fairly easy to depict the trajectory of the stock. However, several studies using this task and other stocks and flows based problems found that the average performance in these tasks was less than 50 % (Sweeney and Sterman 2000; Moxnes 2004; Cronin et al. 2009; Qudrat-Ullah 2014). This empirical evidence is illustrates the power of understating the basic dynamics of complex systems in terms of stocks and flows. Understanding the nature and dynamics of an energy system's stocks and flows, therefore, is an essential ingredient of an effective and efficient energy policy of a nation or region.

The primary purpose of this book, therefore, is to aid energy policy makers to understand better the dynamics of interconnected, sociotechnical stocks and flows of energy systems, whereby often well-intentioned and apparently sound energy polices produce counterintuitive outcomes. By using the system dynamics approach (Forrester 1961), decision makers can explicitly see the representation and impact of energy systems' stakeholders' point of view (e.g., economics, environmental, geopolitical, and ethical). It is not the question of only supply and demand of energy resources but how we can best manage these limited resources subject to local (e.g., adequate, affordable, and cleaner availability of energy) and global (e.g., regulations, politics, and security) constraints. Poor assumptions adversely impact even the well-intentioned decisions (e.g., to see how poor assumptions, e.g., technology challenges, have been seriously underestimated even by the Intergovernmental Panel on Climate Change, please see Pielke et al. 2008). The ability of the system dynamics approach to allow the decision makers to test their assumptions and develop consensus-oriented energy policies is the currency of this approach.

Therefore, to achieve the objective of this book, "to enhance systematically our understanding of and gain insights into the general process by which the physics of stocks and flows of an energy system explains the making of an efficient and effective energy policy of a nation," our focus is on understanding of the dynamics of key stocks and flows of energy systems. In this brief, we identify some fundamental accumulation processes (stocks), spanning across the demand and supply sectors of energy systems that are common to the majority of energy policy related issues across the globe. By presenting some examples of system dynamics simulation models wherein these accumulation processes are drivers of the behavior of the represented energy system, we demonstrate the importance and utility of these fundamental and foundational structures of energy systems. Based on these modeling efforts, three cases, (i) socioeconomic and environmental implications of the energy policy of Pakistan, (ii) the dynamics of green power in Ontario, and (iii) the dynamics of electricity generation capacity of Canada, are analyzed and discussed. We begin our journey by presenting, in the next section of this chapter, a general overview of the dynamics of an energy policy.

1.2 Dynamics of Energy Policy: An Overview

People from all walks of life regardless of their origin, creed, sect, social strata, and value systems desire a better life. Energy plays a fundamental role in the well-being, prosperity, and improved lives of all of us. Planners and decision makers of any region, state, or country, therefore, have or at least aspire to have a common goal: an adequate, reliable, cleaner, and affordable supply of energy. In order to achieve this noble goal, the responsible governments and authorities, often at the time of the promulgation of their national development plans and economic agendas, enact and implement, assess and adjust various energy policies, often called "national energy policy." For instance, the president of the United Sates stated the objectives of the national energy policy in 2001 as (NEP 2001):

> to develop a national energy policy designed to help the private sector, and, as necessary and appropriate, State and local governments, promote dependable, affordable, and environmentally sound production and distribution of energy for the future.

The Government of Ontario, Canada's largest province both by population and by electricity generation capacity promulgated its 2013 Long-Term Energy Plan with the objective:

> to balance the following five principles: cost-effectiveness, reliability, clean energy, community engagement and an emphasis on conservation and demand management before building new generation.

To realize such governmental policy objectives, the common challenge for the decision makers is how to design an effective and efficient energy policy subject to sociotechnical constraints. To begin with, the design of an effective and efficient energy policy is a complex task. What kind of complexity do the decision makers of energy policy face?

To be fair, the concept of complexity means something different things to different people. However, there are two main types of complexity: combinatorial complexity (or detail) and dynamic complexity. Although combinatorial complexity stems from a fairly large number of variables and even a larger number of possible combinations of these variables of a system (e.g., scheduling of games of the world soccer league or scheduling of exams in a short exam period of a large university), dynamic complexity arises due to the interactions of actors of a system over time. In fact, Sterman in his pioneering work (2000, p. 21) elegantly describes the distinctive feature of dynamic complexity as

> Most people think of complexity in terms of the number of components in a system or the number of combinations one must consider in making a decision. The problem of optimally scheduling an airline's flights and crews is highly complex, but the complexity lies in finding the best solution out of an astronomical number of possibilities. Such needle-in-a-haystack problems have high levels of combinatorial complexity (also known as detail complexity). Dynamic complexity, in contrast, can arise even in simple systems with low combinatorial complexity. Dynamic complexity arises from the interactions of the agents over time.

Therefore, the task of making an effective and efficient energy policy is a dynamically complex task. In such tasks, causes and effects are often distant in time and space: making policy decision making a difficult task.

Treating the task of energy policy design as a purely technical task won't be of much help in achieving the objective of an energy policy either. Energy systems are essentially sociotechnical systems. The complexities of physical energy generation systems (e.g., run-of-river and reservoir-based hydro power plants, nuclear power plants, wind, solar, biomass, and tidal power generation technologies, transmission and distribution systems) are highly technical, however, the nature of issues related to affordability (e.g., ability of consumers to pay) and adoption of technologies (e.g., role of consumers' sensitivity to environmental emissions) is predominantly social. The limited resources and conflicting objectives of various stakeholders (e.g., local community, business community, regulatory agencies, and governments) further complicate the energy policy decisions. Figure 1.1 presents an example of the dynamics of a few variables relevant to the energy system of Canada. Here, the variable Oil Production represents the technical aspect of the energy system (e.g., drilling capacity and efficiency are important indicators of the oil production capability of an energy system). On the other hand, population growth and energy prices are socioeconomic factors of an energy system. Expensive energy badly affects the budget of low-income nations across the globe. Environmental pollution severely impacts the well-being of people. Perceived risks of nuclear power are increasingly high. Building of large dams for hydro power has detrimental implications for the habitat. The intertwined nature of these socioeconomic and environmental factors of an energy system adds to the complexity of the task of energy policy decision makers. Even a systematic visual look at the varying behaviors of the selected four variables (Fig. 1.1): Natural Gas Prices, Electricity Prices, Oil Production, and Population Growth Rate, reveals a critical reality pertaining to the management of energy systems. That is, to manage an energy system effectively, the use of traditional statistical estimation alone (e.g., regressions and trending) will hardly be of any help. Instead, an appreciation and understanding of the dynamics of these sociotechnical systems is an essential prerequisite of the design of an energy policy.

In complex dynamic systems such as an energy system, stocks or the accumulation processes of the system are a major force behind the dynamic behavior of the system (Forrestor 1961; Sterman 2000). For instance, new investments made in the

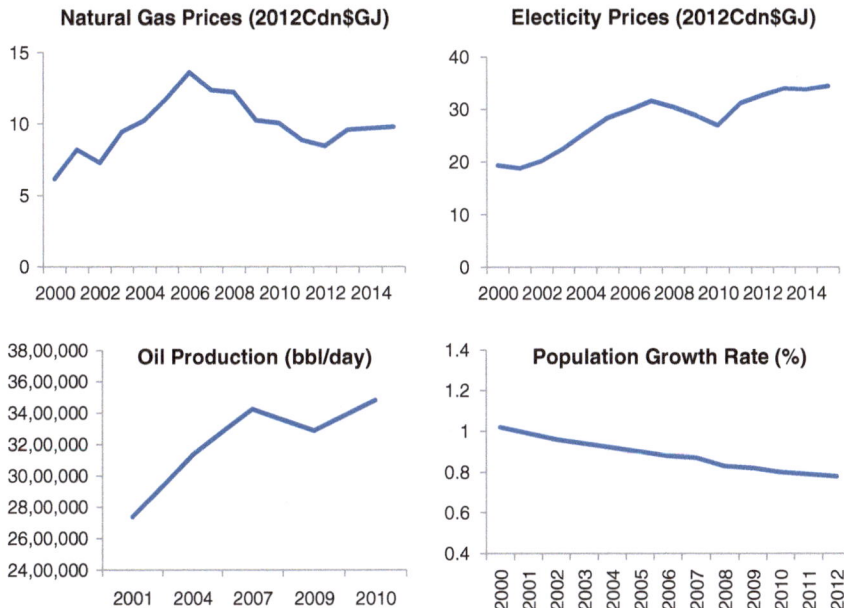

Fig. 1.1 Canadian population growth rate, oil production, and prices of gas and electricity. *Note* 2012Cdn$: 2012 Canadian Dollar; GJ, bbl: Giga Joul billion barrel

electricity generation capacity lead to an increase in the stock of "generation capacity". This stock of "generation capacity" can decease upon aging (i.e., a power plant becomes obsolete when its operational life expires). This stock or the accumulation process will exhibit a dynamic equilibrium when net flow (i.e., addition of capacity minus deletion of capacity) is zero. Likewise, population dynamics (e.g., new births, deaths, and different strata of population) impact the demand of electricity. So, to understand the demand of electricity better, one should appreciate and understand the dynamics of the underlying stock, i.e., the population. Knowledge about the physics of stocks (i.e., the dynamic behavior) of a complex system, that a stock can only be controlled through its inflows and outflows, plays a key role in the understanding of the dynamics of that complex system. Therefore, a better understanding of the physics of stocks and flows of a complex dynamic system is required for the effective and efficient management of such a system, a raison d'être of any energy policy.

1.3 Organization of This Book

This book is organized into three major parts. Part I (i.e., This chapter and Chap. 2) (i) serves as an introduction to the subject of "modeling for energy policy," its importance and complexity, and (ii) introduces the system dynamics approach as a "solution" modeling approach. Part II (i.e., Chaps. 3, 4 and 5) focuses on (i) the physics of stocks and flows and how it deals with the complexity of energy systems and (ii) the validation procedures for the stock and flow based simulation models. Finally, Part III (i.e., Chaps. 6 and 7) presents the applications of system dynamics models developed for dealing with various energy policy issues, thereby helping driving critical lessons to be learned.

1.4 Summary

All energy policies, be it for a developing or a developed region, are designed to meet the socioeconomic objectives set forth by the decision makers. The dynamic and intertwined nature of socioeconomic and environmental factors of an energy system adds to the complexity of the task. Nevertheless, all the modeling efforts regarding the stocks and flows oriented modeling endeavors for energy systems ought to be geared to achieve those policy objectives.

Identification of the key stocks and flows of an energy system and the understanding of the physics of these interacting stocks and flows play a fundamental role in the design and assessment of an effective and efficient energy policy. In the next chapter, we present an overview of the practice of simulation and modeling in service of energy policy and related issues.

References

Cronin, M., Gonzalez, C., & Sterman, J. (2009). Why don't well-educated adults understand accumulation? A challenge to researchers, educators, and citizens. *Organizational Behavior and Human Decision Processes, 108*, 116–130.

Forrester, J. (1961). *Industrial dynamics*. Cambridge, USA: MIT Press.

Moxnes, E. (2004). Misperception of basic dynamics: The case of renewable resource management. *System Dynamics Review, 20*(2), 139–162.

NEP. (2001). National Energy Policy. Report of the Nation Energy Policy Development Group. Washington, DC: US Government Printing Office. (ISBN 0-16-050814-2). Sterman, 2000.

Pielke, R, Jr, Wigley, T., & Green, C. (2008). Dangerous assumptions. *Nature, 452*, 431–432.

Qudrat-Ullah, H. (2014). *Better decision making in complex, dynamics tasks*. New York, USA: Springer.

Sterman, J. (2000). *Business dynamics: systems thinking and modeling for a complex world*. NY, USA: McGraw-Hill.

Sterman, J. D., & Sweeney, L. (2002). Cloudy skies: Assessing public understanding of global warming. *System Dynamics Review, 18*(2), 207–240.

Sweeney, L., & Sterman, J. (2000). Bathtub dynamics: Initial results of a systems thinking inventory. *System Dynamics Review, 16*(4), 249–286.

Chapter 2
Modeling and Simulation in Service of Energy Policy: The Challenges

People would rather believe than know
—Edward O. Wilson

Modeling and simulation have long and well served the actors and various decision makers in the domain of energy policy. Various modeling approaches and models have been applied to address a variety of energy policy related issues. However, the journey continues. This chapter provides an overview of these modeling approaches and models identifying their key challenges in the face of emerging issues. The identified energy policy modeling related issues include the characterization of energy systems as complex dynamic systems with numerous uncertainties, nonlinearities, time lags, and intertwined feedback loops.

2.1 Energy Systems Modeling and Its Challenges

By and large the modeling and simulation community has successfully used a variety of methods and techniques to serve energy policy needs. For instance,

- *Linear programming and dynamic programing:* to perform capacity expansion and energy-economy analysis [e.g., WASP model (Foel 1985), MARKAL model (Fishbone and Abilock 1981), and RES model (Howells et al. 2011)]
- *Mixed-integer linear program*: to optimize distributed energy resource system [e.g., MILP model (Omu et al. 2013)]
- *Econometric methods:* to produce annual energy outlook and the role of carbon capture and storage [NEMS model (Kydes and Shah 1997) and SGM model (Praetorius and Schumacher 2009)]
- *Partial equilibrium model*: to develop the US Climate Action Plan [e.g., IDEAS model (Wood and Geinzer 1997)]
- *Optimization*: to analyze energy–economy interactions and optimize the options for SO_2 control (e.g., Meier and Mubayi's (1983) model and Islas and Grande's (2007) model)

© The Author(s) 2016
H. Qudrat-Ullah, *The Physics of Stocks and Flows of Energy Systems*,
SpringerBriefs in Complexity, DOI 10.1007/978-3-319-24829-5_2

- *Scenario analysis*: to analyze energy policies (e.g., Munasinghe and Meier's (1993) model
- *Agent-based modeling*: to provide quantitative support for climate policy formulation and evaluation [e.g., ENGAGE model (Wang et al. 2013)]

have been applied to address various energy policy related issues, be it in a developing or a developed region or country. Despite the demonstrated applicability and success of these operational methods over the past several decades (Dyner and Larsen 2001), emerging issues related to the energy industry (e.g., widespread deregulated electricity markets and industry, climate change and environmental concerns, multiple stakeholders, and technological disruptions) require new capabilities of modeling methods to fully capture the dynamics of energy systems. These energy system modeling[1] challenges include modeling the existence of uncertainties (e.g., fuel prices), time delays (e.g., power plant construction time lags), nonlinear causal relationships (e.g., between changes in electricity price and its industrial use), and interacting feedback loops (e.g., additional production capacity brings in more revenue for the firm, which, in turn, leads to increased production capacity) in any energy system.

2.1.1 Uncertainties Abound

In general, widespread deregulation and privatization in the energy sector of the economies has created opportunities as well as challenges for private investors including independent power producers (IPPs). In the case of developing and emerging nations including India, China, and Brazil, growing demand and consumption of energy create imbalances providing further impetus for energy sector investments (IEA 2012). However, the dynamics of the much desired stock of "investments" in the energy sector are uncertain:

i. *The nature and life of incentives and rules keep changing.* Although the learning aspects of these changes are desired, the resulting often costly, lengthy, and uncertain litigations deter potential new investments in the energy sector of the host country (e.g., Eberhard and Gratwick 2007).
ii. *Technological disruptions can severely impact investments.* Costly retrofitting or installation of new technologies, say for monitoring and control of electricity production related environmental emissions is becoming common and is highly unpredictable.
iii. *The availability and prices of fuels are rarely in smooth order.* This creates operational and financial difficulties for the energy projects.

---

[1]The basic premise of modeling a system is that we are able to identify the forces (i.e., components or structures of a system) behind the problematic behavior of this system (e.g., the supply of electricity lags or leads the demand). By controlling/managing the underlying structures of a system, the problematic behavior can be improved.

iv. *Deregulation has expanded the nature and dimensions of stakeholders.* Compared with the almost monopolized status of regulated regimes, now multiple stakeholders including competitors, regulators, traders, large institutional investors, shareholders, local communities, end-users, and environment lobbyists are involved in energy sector investments. Not only are they "many more" but these stakeholders often come with conflicting objectives, making energy policy decisions even more complex.

v. *Perceptions of people change and sometimes in relatively short order.* For instance, after the Fukushima nuclear accident in Japan in 2011, Germany and Switzerland decided relatively quickly to close their nuclear power plants (Larsen and Arrango 2013). Granted that the unpredictability of such external events is known, the ability of decision makers to conceive and explore such scenarios can better prepare them to deal with such uncertainties (Wang et al. 2013).

2.1.2   Existence of Nonlinear Relationships Is a Reality

In energy systems, there exist nonlinear relationships between variables of the system that can hardly be analyzed with traditional econometric methods and linear programming techniques. For instance, when the price of electricity decreases, its industrial usage can see some growth (as is shown in Fig. 2.1). However, after a while, when even the price continues to fall, industrial usage of electricity will saturate (e.g., because the production reaches its maximum capacity). Likewise, the relationship between an operator's overtime work and her productivity is nonlinear; in the beginning, her productivity can increase (e.g., due to learning) but if she continues to overwork for long then her productivity will fall or even complete collapse, the *burn-out phenomenon*, can occur. Productivity gained by experienced power plant operators rarely follows a proportional path: more experience leads to

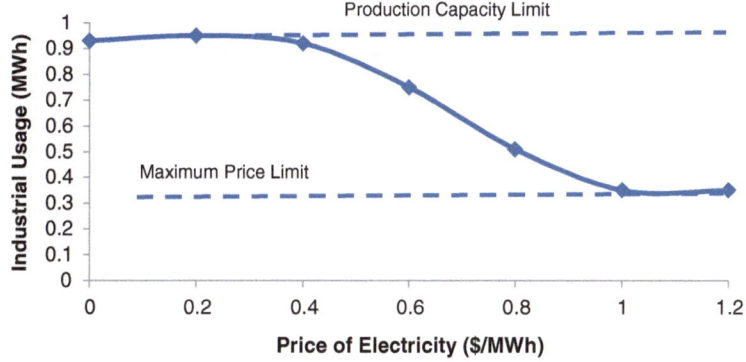

Fig. 2.1 Nonlinear relationship between industrial usage and price of electricity

increased productivity but after some time productivity reaches a plateau. Such nonlinear relationships abound in sociotechnical systems such as energy systems. Therefore, the utility of policy-supporting analysis of an energy system but without an explicit representation and modeling of its critical nonlinear relationships is limited, at best.

2.1.3 Time Lags Can't Be Ignored

Delays are inherent in energy systems. Consider the case of a new investor in, say a gas-fired power plant. The major milestones of this new project, including approval of the application, securing project funding, construction, testing, and commissioning of the power plant, not only take time but often are characterized by delays. In general, these delays are of two kinds: (i) material delays [e.g., delay in the construction of power plants; e.g., on average it takes 3–4 years to build thermal power plants and 6–10 years for nuclear and hydro (reservoir-based) power plants (IAEA 1993)], and (ii) information delays (e.g., delay in the notice of approval of the application and commissioning permit, etc.). These delays have severe implications not only for the power plant investors themselves (e.g., delayed operations mean much delayed earnings leading to, say shareholders' discomfort) but also for the relevant energy planners and decision makers (e.g., the concerns of off-the-grid industry and population). Therefore, the modeling method for energy policy should have the capability to account for the potential dynamics of these inherent time lags of energy systems.

2.1.4 Causation Not Correlation Informs Strategic Decisions

Indeed energy policy decisions are strategic decisions: these decisions dictate the nature of the future energy supply mix and influence the associated economics for the region. It is the information about the causal nature of the relationships between the variables of the energy systems that is useful for enacting an integrated energy policy. For instance, energy policy makers are interested in knowing the influence (s) of the various stocks of the energy system, such as how the stock of "electricity capital" (i.e., various power plants) impacts electricity prices over time. Or which electricity supply mix can provide affordable and cleaner electricity? What would be the long-term impact of certain policy regulations and incentives? Therefore, the candidate modeling method for energy policy should not only be able to represent such causal relationships in the model but also provide information on the dynamics of these influences.

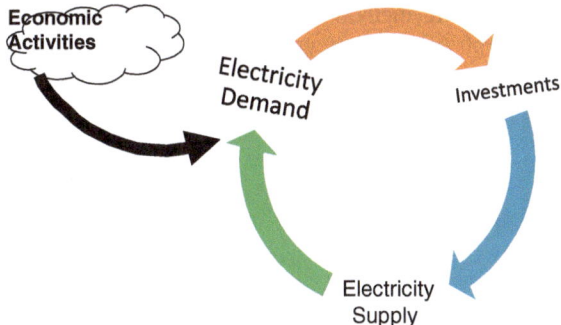

Fig. 2.2 A demand–supply balancing feedback loop

2.1.5   Energy Systems Are Essentially Feedback Systems

Increased economic activities lead to higher electricity demand. Higher electricity demand requires new investments. New investments, after some delays, provide more electricity to close the loop (i.e., either the demand is fulfilled or the cycle, demand → investments → supply → demand, continues until the demand is fully met). Such a cycle is essentially a feedback loop where three variables of an energy system, demand, investments, and supply, are responsible for the resulting dynamic behavior of this feedback loop (as is shown in Fig. 2.2).

There exist several such feedback loops in an energy system and they are interact with each other to produce the dynamic behavior of the energy system (e.g., a particular trajectory of electricity prices, environmental emissions, the stock of renewable technologies (e.g., windmills), sector-related employment, etc.), much needed information for the decision makers to enact a systematic and integrated energy policy. Traditional modeling approaches are hardly adequate for providing such a feedback-oriented analysis of the energy systems to the energy policy decision makers.

2.2   Summary

Overall, modeling and simulation have well served the energy domain for well several decades.

The existence of nonlinear and uncertainty intensive variables, several inherent time lags, and intertwined feedback loops in an energy system pose serious modeling challenges. Now, the increasing liberalization and privatization, heightened emphasis on environmental issues including global warming and climate change, complexity of multidimensional and conflicting interests of stakeholders, and unprecedented technological disruptions have only added to the complexity of the task for energy policy decision makers across the globe. In the context of these forceful developments, the traditional econometric methods and linear programming methods alone are not adequate to deal with the complex dynamic nature of energy policy issues. How do we deal with such complex systems? System dynamics methodology (Forrester 1961) rises to this challenge. Chapter 3, provides details on this promising assertion.

References

Dyner, I., & Larsen, E. R. (2001). From planning to strategy in the electricity industry. *Energy Policy, 29*(13), 1145–1154.

Eberhard, A., & Gratwick, K. (2007). From state to market and back again: Egypt's experiment with independent power projects. Working Paper, Oct 2007. http://www.gsb.uct.ac.za/files/Egypt_IPP_Experience_April_2006.pdf. Accessed on January 8, 2015.

Fishbone, L. G., & Abilock, H. (1981). MARKAL, A linear programming model for energy systems analysis: Technical description of the BNL version. *International Journal of Energy Research, 5,* 353–375.

Foell, W. K. (1985). Energy planning in developing-countries. *Energy Policy, 13,* 350–354.

Forrester, J. (1961). *Industrial dynamics.* Cambridge, USA: MIT Press.

Howells, et al. (2011). OSeMOSYS: The open source energy modeling system: An introduction to its ethos, structure and development. *Energy Policy, 39,* 5850–5890.

IAEA. (1993). Energy and Nuclear Power Planning study for Pakistan: 1993-2023. IAEA-TECDOC-1030, IAEA, Vienna, Austria, www-pub.iaea.org/MTCD/Publications/PDF/te_1030_prn.pdf. Accessed on November 11, 2014.

IEA. (2012). IEA report on sustainable energy for all. V2. https://www.iea.org/media/freepublications/oneoff/GlobalTrackingFrameworkOverview.pdf. Accessed on January 17, 2015.

Islas, J., & Grande, G. (2007). Optimization of alternative options for SO_2 emissions control in the Mexican electrical sector. *Energy Policy, 35*(9), 4495–4503.

Kydes, A. S., & Shah, S. H. (1997). The national energy modelling system: Policy analysis and forecasting at the US department of energy. In D. W. Bunn & E. R. Larsen (Eds.), *Systems modelling for energy policy* (pp. 9–30). Chichester, England: Wiley.

Larsen, R., & Arango, S. (2013). System dynamics and the process of electricity deregulation. In H. Qudrat-Ullah (Ed.), *Energy policy modeling in the 21st century* (pp. 17–30). NY, USA: Springer.

Meier, P., & Mubayi, V. (1983). Modeling energy economic interactions in developing-countries: A linear-programming approach. *European Journal of Operational Research, 13,* 41–59.

Munasinghe, M., & Meier, P. (1993). *Energy policy analysis and modeling.* Cambridge: England.

Omu, A., Choudhary, C., & Boies, A. (2013). Distributed energy resource system optimization using mixed integer linear programming. *Energy Policy, 61,* 2490266.

Praetorius, B., & Schumacher, K. (2009). Greenhouse gas mitigation in a carbon constrained world: The role of carbon capture and storage. *Energy Policy, 37,* 5081–5093.

Wang, P., Gerst, D., & Borsuk, E. (2013). Exploring energy and economics futures using agent-based modeling and scenario discovery. In H. Qudrat-Ullah (Ed.), *Energy policy modeling in the 21st century* (pp. 251–269). NY, USA: Springer.

Wood, F. P., & Geinzer, J. C. (1997). The IDEAS model and its use in developing the US climate change action plan. In D. W. Bunn & E. R. Larsen (Eds.), *Systems modelling for energy policy* (pp. 31–46). Chichester, England: Wiley.

Chapter 3
Meeting the Challenges: Energy Policy Modeling with System Dynamics

> The only way to find out what will happen when a complex system is disturbed is to disturb the system, not merely to observe it passively
>
> —George Box

Given that energy systems are feedback systems, system dynamics methodology appears to be a natural choice for the energy policy modeling community. In fact, system dynamics methodology has long served the energy policy domain. Among its strengths is that, it provides a powerful language to represent the causal relationship among the variables of an energy system. First, this chapter provides a brief overview of how system dynamics has been applied to a variety of issues related to the energy policy area. Second, to explain the fundamental structures of energy systems, feedback loops, and the causal loop diagram (CLD) are introduced. Procedures and rules on a feedback loop polarity are elaborated. The utility of a CLD in the construction of a dynamic hypothesis of complex problems is also discussed. As there is no feedback without the stocks and flows of a system, the language of the stocks and flows is introduced as well.

3.1 System Dynamics and Energy Policy Modeling

The system dynamics (SD) approach (Forrester 1961) takes a feedback perspective to describe and explain the dynamics (i.e., changes over time) of complex systems. The fundamental premise of system dynamics is that the structure of a system drives its dynamic behavior, the endogenous view (Sterman 2000). For instance, fluctuations in the price of crude oil (behavior of the oil supply–demand system) are due to internal factors (e.g., disruptions in oil production, delays in the transportation of oil: internal structures of the oil supply–demand system). Thus, as per the SD view, it is imperative that a better understanding and control of the (internal)

H. Qudrat-Ullah, *The Physics of Stocks and Flows of Energy Systems*, SpringerBriefs in Complexity, DOI 10.1007/978-3-319-24829-5_3 structures of an energy system should lead to better (behaviors) outcomes (e.g., adequate and affordable supply of electricity). Likewise, a utility company's loss of market share, according to an SD perspective, is due to internal (to the company) factors (e.g., low level of investments and low innovation rate in customer service sector, the stocks and flows of the system).

In fact, system dynamics models have been serving the energy policy domain for more than five decades. For instance, in 1973, Fossil, a system dynamics model, was developed in the United States for energy policy design, with the explicit focus on all the sources of energy and energy demand (Naill 1973). Building on the successes of Fossil, later on the IDEAS model was developed and was the official Department of Energy's energy planning model for the United States until 1995. The energy demand; electricity generation; oil, gas, and coal production; renewables; and environmental emissions were explicitly modeled in IDEAS (Amlin 2013). Ford's work on the Pacific Northwest Hydroelectric System and electric utility is a solid body of system dynamics modeling (for an excellent review of Ford's work and other electricity-related system dynamics modeling and applications, please see Ford 1996). Adding the climate change context to the existing energy–economy interactions focused system dynamic models, FREE is a model that tests the implications of climate change related feedback processes (Fiddaman 2002). Another well-known system dynamics model, ENERGY 2020, has been actively used by regional and national governments to conduct energy and emission related policy analysis in the United States and Canada (Amlin 2013).

In the early 1990s, increased deregulation and liberalization in the energy sector of various countries introduced new uncertainties for energy planners and decision makers, giving rise to an even higher demand for system dynamics models. System dynamics models were applied to several energy policies and related issues. The prominent system dynamics models for various energy policy related themes are listed in Table 3.1. As expected, much of the modeling activity took place in the context of regional and national energy policies, liberalization and privatization of electricity systems, and environmental concerns.

In all of these system dynamics models, various stocks and flows structures are the foundational elements. Instead of going through lists of these structures (interested readers can access these models through the listed references), we are focusing on some common key stocks and flows that in general are responsible for the dynamic behavior of an electricity system. Figure 3.1 provides a macrolevel view of these fundamental structures and the potential interactions among them.

In fact, in Fig. 3.1, there are several feedback loops in action. For instance, the demand for electricity—based on regulation and incentives—will generate investments for production capital. The production capital, once fully operational, with the availability of the required labor and resources (i.e., fuels) will produce the

Table 3.1 System dynamic energy policy models

Energy policy modeling themes	Sources
• Design and assessment of regional and national energy and electricity policies	Naill (1973), Davidsen et al. (1990), Dyner and Bunn (1997), Qudrat-Ullah and Davidsen (2001), Pasaoglu and Or (2006), Park et al. (2007), Dyner et al. (2009), Van Ackere and Ochoa (2009), Bassi et al. (2013), Amlin (2013), Aslani et al. (2014), Qudrat-Ullah (2015)
• The effects of privatization of electricity and firms' investment behavior across generation technologies	Bunn and Larsen (1992), Bunn et al. (1993), Bunn and Larsen (1994, 1999), Dimitrovski et al. (2007a)
• Energy efficiency analysis and management	Dyner et al. (1995), Qudrat-Ullah (2013)
• Electricity market design and responses	Vlahos (1998), Dyner (2001), Ochoa (2007), Dyner et al. (2009), Arango and Larsen (2011)
• Generation capacity expansion	Ford (2001), Olsina et al. (2006), Dimitrovski et al. (2007b), Hasani and Hosseini (2011), Qudrat-Ullah (2013)
• Renewables and environmental emissions	Vogstad et al. (2002), Anand et al. (2005), Fiddaman (2002), Jin et al. (2009), Ford et al. (2007), Han and Hayashi (2008), Trappey et al. (2012), Qudrat-Ullah (2005, 2014), Feng et al. (2013), Saeed (2013)

These 37 publications are not meant as an exhaustive list of system dynamics work on energy policy modeling but fairly representative work on various thematic areas. The interested reader can see a comprehensive account of system dynamics modeling work in the energy domain at: http://www.iip.kit.edu/downloads/1_Teufel_Review_of_Electricity_Models_with_System_Dynamics.pdf

necessary electricity to meet the demand, bringing a closure to this demand–investments–supply feedback loop. On the other hand, consider the electricity intensity. The higher the electricity intensity, the higher the demand will be, which in turn will further increase the intensity and the vicious cycle (i.e., a progrowth feedback loop) continues. Another feedback loop counterbalances this growth loop: with the increased conservation and efficiency (e.g., of the appliances), there will be a relatively lower demand for electricity. Lower demand will lead to lower electricity intensity and this virtuous cycle will continue until the electricity system is balanced; the demand and supply of electricity is matched subject to the socioeconomic and environmental constraints (please see Fig. 2.2, where there are several feedback loops involving the stocks and flows of the environment and socioeconomic sectors). These interacting feedback loops constitute a web of fundamental structures, which are responsible for the generation of dynamic behavior of the complex energy system.

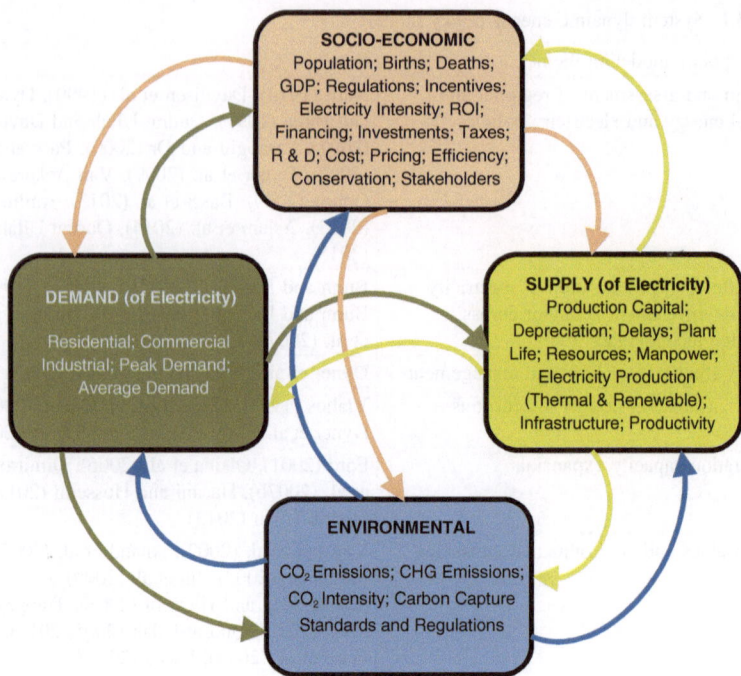

Fig. 3.1 A feedback-oriented model of an electricity system

3.2 Modeling Causality in Systems Dynamics

Given that energy systems are feedback systems, to better understand an energy system requires the appreciation and understanding of its structures and their dynamics. As a first step towards our understanding of an energy system, therefore, we need to describe it: what its key variables are and how they are related to each other in feedback loops. System dynamics provides a powerful language to describe the causal relationships among the variables of the feedback loops of a system.

Cause–effect relationships: Consider the two fundamental variables of any energy system: demand and supply. In system dynamics, the cause–effect relationship, changes in demand of electricity cause changes in electricity supply (in the same direction), is represented as (the positive sign at the head of the arrow means that if the variable at the tail is increased, so will be the variable at the head; if the variable at the tail is decreased, the variable at the head will also decrease) a positive cause–effect relationship:

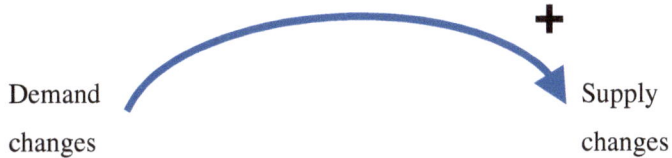

Demand
changes

Supply
changes

On the other hand, the causal relationship between conservation and demand of electricity, changes in conservation cause changes in demand for electricity (in the opposite direction, an increase in conservation will cause a decrease in demand for electricity) is represented by placing a negative sign at the head of the arrow between the two variables, a negative cause–effect relationship:

Conservation
changes

Demand
changes

Time lags: There are several time delays (i.e., time lag between actions and their consequences) among variables of an energy system. The cause–effect relationship between demand and supply, as represented above, has a delay between both variables. For instance, if the demand for electricity increases (e.g., due to increased industrial activity), only after a substantial delay (e.g., due to the construction of new power plants to meet this additional demand), the additional supply of electricity can be operationalized. There are two types of delay, information delay (e.g., it takes time to get approval from authorities before construction of a power plant can begin) and material delay (e.g., time it takes to move fuel oil from a central storage facility to the site of the power plant). Such delays are represented by having a symbol ‖ on the arrow that links the two variables, as

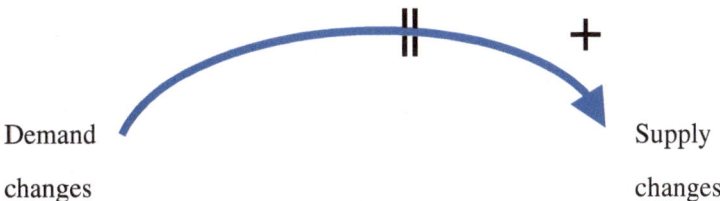

Demand
changes

Supply
changes

Fig. 3.2 A demand–supply
balancing feedback loop

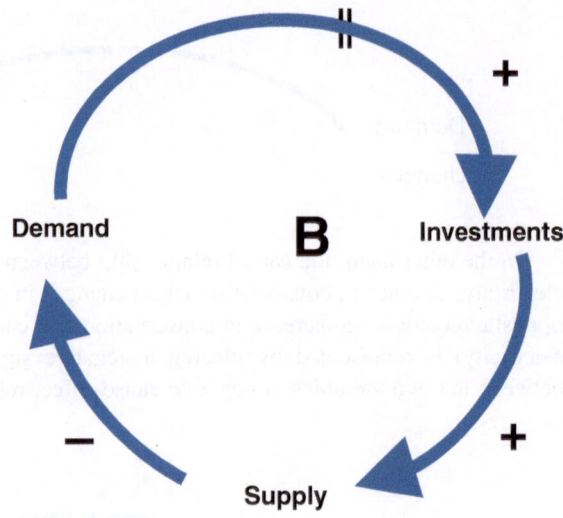

Feedback loops: As realized earlier, an energy system is essentially a feedback
system; it has several interconnected feedback loops. A feedback loop is a closed
sequence of various cause–effect relationships. For instance, Fig. 3.2 presents a
demand–investments–supply feedback loop: three cause–effect relationships
forming a closed sequence. Now if we begin with a change (increase or decrease) in
any of these three variables and trace the direction of this change around the loop,
we will come back to the same variable but with a change in an opposite direction.
For instance, an increased demand (positive direction of change) leads to increased
investments in the electricity generation technologies. In turn, increased investments, after a delay, cause an increase in the supply of electricity. The increased
supply then will close or reduce the gap between demand and supply or result in a
decreased demand of electricity; we started with increase in demand but closed the
loop with a decrease in demand. Likewise, if we start the loop with a decrease in the
demand of electricity, we will end up the loop with an increased demand of electricity: a decreased demand requires relatively decreased new investments; lower
investments will cause a decrease in the supply of electricity. The decreased supply
will widen or increase the gap between demand and supply or result in a relatively
increased demand for electricity. Such a feedback loop is called a negative or
balancing loop, **B**. In this balancing loop, there exist two delays, one between the
Investments and the Supply and the other between the Demand and the
Investments. These delays cause instability in the dynamic behavior of any feedback loop (Sterman 2000).

Figure 3.3 represents another kind of feedback loop, a positive or reinforcing
feedback loop, **R**. Here, say we start with an increase in the supply of
electricityElectricity by the producer, improving its profitability. The increased
profits will cause more investments, eventually improving its profitability even
more, a growth-propelling loop.

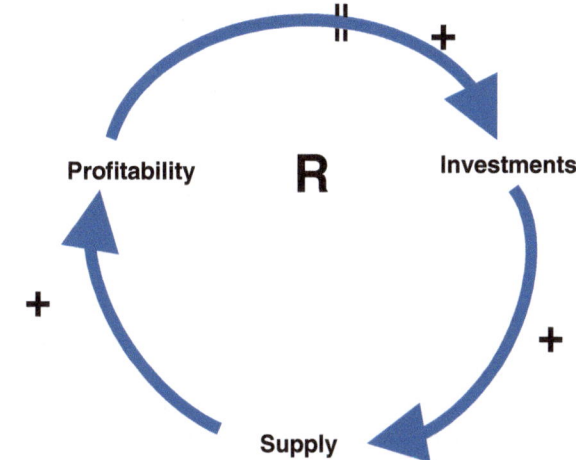

Fig. 3.3 A profitability–supply reinforcing feedback loop

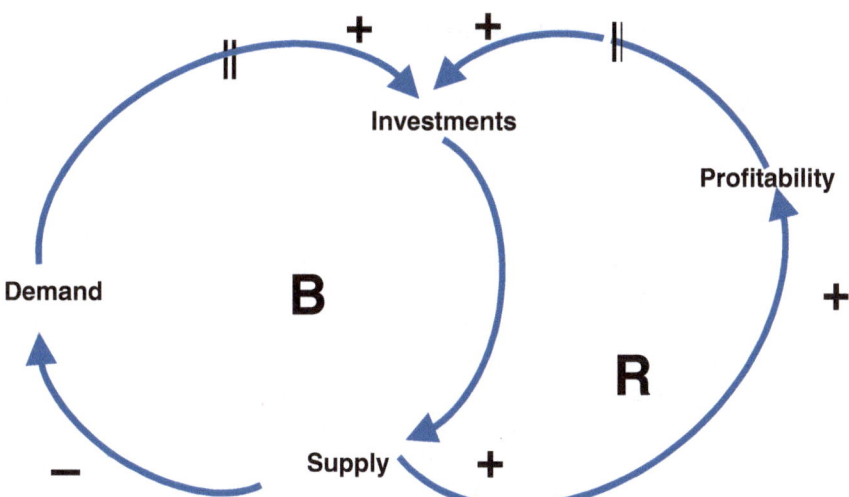

Fig. 3.4 A CLD of a simple dynamic model

In system dynamics, if we combine Figs. 3.3 and 3.4, the resulting figure is called a causal loop diagram, a diagram that visually represents the feedback loops of a system (e.g., Fig. 3.4). Figure 3.4, a CLD, represents two feedback loops of a system, one positive and the other negative. How do we know whether a feedback loop is positive or negative? There is a simple, two-step visual counting method:

Step 1:   Look at all of causal links of a feedback loop
Step 2:   Count the negative (−) signs

If the total number of negative signs is zero (i.e., all signs are positive) or even
        Then the loop polarity is positive (i.e., the feedback loop is positive)
                Else (i.e., if negative signs are odd, e.g., 1, 3, 5, etc.) the loop polarity is
                        negative (i.e., the feedback loop is negative)

When it comes to policy design, the involvement of stakeholders is a critical step. CLDs provide a powerful technique to facilitate dialogue and build the consensus on the key actors and processes of their energy system (Bassi et al. 2013). In this initial phase of modeling, stakeholders' participation (e.g., higher management) can achieve better support and ownership by management (Vennix, 1996), a critical factor in the successful implementation of the mode-based policy insights and recommendations later on.

3.3 CLD as a Tool to Represent a Dynamic Hypothesis

Before building a formal model of stocks and flows of a system, a CLD represents the conceptual model of the underlying system. In system dynamics, often such a conceptual model is called a dynamic hypothesis. A dynamic hypothesis is a statement about the structure of a system, often represented by a CLD, which is able to generate the problematic behavior of the system (e.g., what causes the imbalance between demand and supply of the electricity system of a region?). Consider Fig. 3.5, a CLD, which represents a dynamic hypothesis: "an environment with higher intensity of CO_2 emissions will be the long-term outcome of the (existing) electricity policy of Pakistan" (Qudrat-Ullah 2014, p. 186).

As depicted in Fig. 3.5, the gap between supply and demand of electricity is shortened by investments being made in electricity-generating technologies (i.e., power plants). These investments will, after a delay, depending on the type of power plant being constructed, result in the electricity-generating capacity (i.e., production capital). The capital together with resources will generate electricity. The storage capacity acts as a limiting factor for onsite resource availability. Once the resource is available at the site, the power plant is ready to generate electricity. Depending upon the type of fuel being consumed, the generation of electricity, except for hydro-based production, will produce carbon emissions.

However, only when environmental restrictions are implemented in the alternative policy, the power plant that emits more CO_2 becomes costly due to the environmental premiums charged. Similarly, if the resource being consumed is imported then an import dependency premium is realized to the cost of the electricity being produced. The more expensive a technology becomes, the less it receives of the share of investments. Also the resource availability impacts the

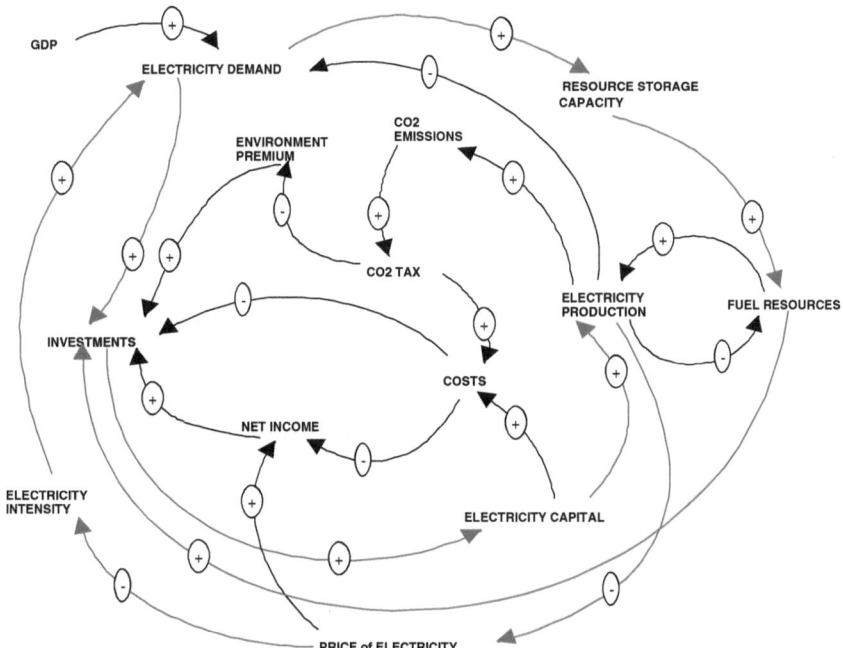

Fig. 3.5 Dynamic hypothesis. *Source* Qudrat-Ullah (2014)

resource price that in turn makes that electricity-generating technology more expensive. In addition to these factors, the standard cost elements including capital cost, fuel cost, and operating cost associated with each of the electricity-generating technologies impact their relative cheapness. Thus, the more expensive a technology, the less is its share in the investments being made by the IPPs (independent power producers). Not only the income alone, but also "how quick an income stream is realized" might also influence the choice of IPPs' investments. The price of the electricity impacts the consumption pattern that is reflected in the electricity intensity. The higher electricity intensity generates the higher demand for electricity that invites the IPPs' investments.

3.4   Energy System as a System of Interconnected Feedback Loops of Stocks and Flows

We assert that all energy systems are a feedback system: a system of interconnected feedback loops (e.g., as shown in Figs. 3.4 and 3.5). In each of these feedback loops, the stocks (e.g., stocks of production capital, investments, fuel resources, population, etc.) and flows (electricity consumption rate, power plant depreciation rate, etc.) of the energy system are present and are primarily responsible for the dynamic behavior of the energy system.

As we have noticed, the CLDs are good at communicating about the cause–effect relationships that exist in a system by explicitly portraying them in various feedback loops. However, a CLD does not distinguish between what flows and where it accumulates, a description of the physics of stocks and flows. In system dynamics, specific conventions are used to represent the stocks and flows of a system in an explicit manner. Section 3.5, provides the details and examples of the representation of stocks and flows.

3.5   The Language: Representation of Stocks[1] and Flows

Consider Fig. 3.6, a stock and flow diagram, a partial model of an energy system model. In this diagram, there are six distinct symbols that are described in Table 3.2. This structure, in terms of system dynamics terminology, consists of one stock, two flows, one auxiliary, and two constants. It describes the accumulation process of electricity production capital. The production capital stock (represented by a rectangle, a Level) increases at a rate (represented by a valve, a Rate) determined by (i) the capital under construction (an auxiliary variable represented by a circle), and (ii) the average construction delay (e.g., how long it will take to build a gas-fired power plant), a constant represented by a diamond. On the other hand, this stock will deplete at a depreciation rate (represented by a valve) which is determined by (i) the existing electricity production capital stock, and (ii) the average life of the capital (e.g., the average life of a nuclear power plant is 40 years), a constant represented by a diamond. It is clear that electricity production capital stock[2] can increase only if the net flow (i.e., outflow minus inflow) is positive (the inflow is greater than the outflow) and vice versa (i.e., the stock will decrease if the net flow is negative).

As the stocks are the accumulation (or integration) of flows, a stock in a feedback loop can only be preceded by a flow. To control a stock or a level (e.g., increase, decrease, or maintain at a specific value), the flows act as the regulators or the decision rules (e.g., what should be the rate of acquisition of the capital?). Should we spend on the refurbishment of the plant to slow the depreciation of the production capital? In the context of energy policy, these are our policy levers: point of interventions (e.g., policy incentives will impact the investment rate that in turn can increase the production capital). Thus, a stock and flow diagram presents a dynamic view of the structural elements of the system whose behavior is of interest to the policy decision makers.

---

[1]The terms *stocks and flows* are synonymous with *levels and rates*, which are commonly used in the physical sciences.

[2]Please note that both the production capital stock and its flows (i.e., capital acquisition rate and capital depreciation rate) are shown in the same color, which is blue in this case. This coloring scheme highlights the fact that whatever material goes into a stock or accumulates (e.g., production capital), the same kind of material outflows from the stock (i.e., apples in, apples out).

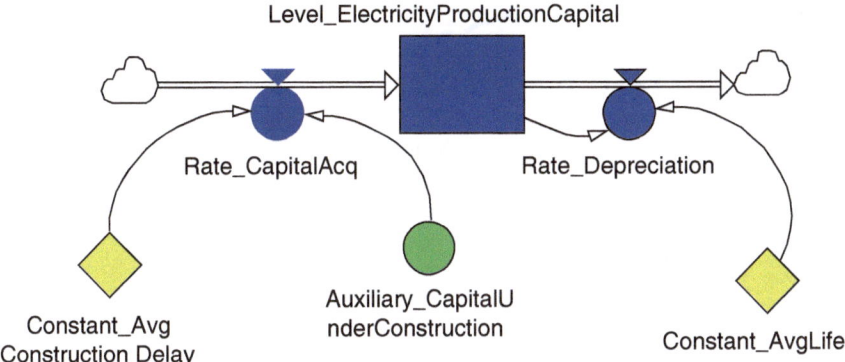

Fig. 3.6 A stock and flow diagram of the accumulation process of production capital

Table 3.2 Nomenclature of a stock and flow diagram

Symbol	Description
	A rectangle: it represents a Level or a stock (in this case a stock of electricity production capital; i.e., power plants)
	A valve (s): a valve represents a Rate or a flow. In this diagram, there are two valves. The attached arrow points to whether this flow is an in-flow (e.g., Rate_CapitalAcq is the capital acquisition rate) or out-flow (e.g., Rate_Depreciation is the capital depreciation rate) to the stock of electricity production capital. It is worth noting that a stock can only be regulated or controlled (e.g., to reach a particular amount through flows or any movement of material or information of a stock depends on its flow(s); i.e., what goes in and what goes out)
	A circle: a circle represents an auxiliary variable. An auxiliary variable (e.g., capital under construction) represents a quantity that is used in determining a flow (e.g., Auxiliary_CapitalUnderConstruction is a quantity (of capital under construction) that together with a constant, Contstant_AvgConstrcutionDelay, determines a flow (i.e., capital acquisition rate)
	A diamond: a diamond is used to represent a contact in the model (e.g., Constant_AvgLife is a constant that represents the average life of a power plant)
	A causal link: it shows the direction of influence of the quantity at the tail towards the quantity at the head (i.e., arrow). For instance, the causal link from the constant, Constant_AvgLife, to the flow, Rate_Depreciation, shows that this constant has an influence on the flow (i.e., this quantity of this rate, in part, depends on the value of this constant)
	A cloud: the symbol of a cloud is used to represent a source or a sink (e.g., in this diagram, we don't model where the depreciated capital goes). Instead, we assume that it goes to an infinite sink

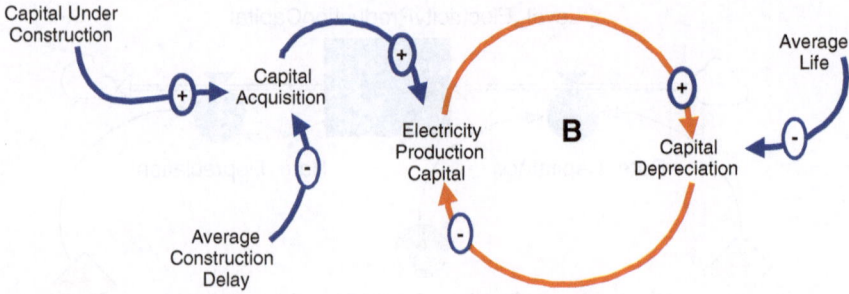

Fig. 3.7 A CLD of the accumulation process of production capital

Now, let us contrast this stock and flow diagram with a corresponding CLD, as given in Fig. 3.7. Here, we can locate a balancing feedback loop, **B**, between a stock variable, *Electricity Production Capital*, and a flow variable, *Capital Depreciation*. Although there are other variables that are causally connected (e.g., Capital Under Construction, an auxiliary variable is connected to a flow variable, Capital Acquisition, and there is a one-way interaction between a flow variable, Capital Acquisition, and a stock variable, Electricity Production Capital) they are not connected through any feedback loop. Therefore, one should note that there is no feedback loop in any system without the existence and a two-way interaction (direct or indirect) of at least one stock variable and one flow variable [for an excellent resource on rules and examples of various stocks and flows, please read Sterman's pioneering work, Business Dynamics (Sterman 2000)].

Using the two diagramming tools of system dynamics, a CLD and a stock and flow diagram, we can develop a causal view of the key structural elements (i.e., feedback loops involving stocks, flows, delays, auxiliary variables, and parameters) of an energy system. However, for an energy policy design and assessment task, we need to understand the dynamics of these structures (i.e., feedback loops). Towards that understanding, the next chapter deals with the physics of stocks and flows.

3.6 Summary

Energy policy design and assessment are complex dynamic tasks. In essence, energy systems are a network of various feedback loops. These interacting feedback loops constitute a web of fundamental structures (e.g., stocks, flows, and delays) responsible for the generation of dynamic behavior of the represented energy system. In contrast to the traditional econometric and statistical methods, system dynamics methodology provides a power language (i.e., CLDs and stocks and flows (SAF) diagrams) to represent causality among the fundamental structures (i.e., feedback loops) of an energy system: a snapshot of the actors and resources of the energy system and who influences what. To better understand the dynamics of these fundamental structures, the next chapter presents methods and techniques to see how these structures contribute to the behavior (i.e., performance) of the energy systems.

References

Amlin, S. (2013). Simulation of greenhouse gas cap-and-trade systems with energy 2020. In H. Qudrat-Ullah (Ed.), *Energy policy modeling in the 21st Century* (pp. 107–122). New York, USA: Springer.

Anand, S., Vrat, P., & Dahiya, R. P. (2005). Application of a system dynamics approach for Assessment and mitigation of CO_2 emissions from the cement industry. *Journal of Environmental Management, 79*, 383–398.

Arango, S., & Larsen, E. (2011). Cycles in deregulated electricity markets: Empirical evidence from two decades. *Energy Policy, 39*(5), 2457–2466.

Aslani, A., Helo, P., & Naaranoja, M. (2014). Role of renewable energy policies in energy dependency in Finland: System dynamics approach. *Applied Energy, 113*, 758–765.

Bassi, M., Deenapanray, P., & Davidsen, P. (2013). Energy policy planning for climate-resilient low-carbon development. In H. Qudrat-Ullah (Ed.), *Energy policy modeling in the 21st century* (pp. 125–156). NY, USA: Springer.

Bunn, D. W., & Larsen, E. R. (1992). Sensitivity of reserve margin to factors influencing investment behaviour in the electricity market of England and wales. *Energy Policy, 20*(5), 420–429.

Bunn, D. W., & Larsen, E. R. (1994). Assessment of the uncertainty in future UK electricity investment using an industry simulation model. *Utilities Policy, 4*(3), 229–236.

Bunn, D. W., & Larsen, E. R. (1999). Deregulation in electricity: Understanding strategic and regulatory risk. *Journal of the Operational Research Society, 50*(4)

Bunn, D. W., Larsen, E., & Vlahos, K. (1993). Complementary modeling approaches for analysing several effects of privatization on electricity investment. *The Journal of the Operational Research Society, 44*(10), 957–971.

Davidsen, P., Sterman, J., & Richardson, G. (1990). A petroleum life cycle model for the United States with Endogenous Technology. *Exploration, Recovery and Demand, System Dynamics Review, 6*(1), 66–93.

Dimitrovski, A., Ford, A., & Tomsovic, K. (2007a). An interdisciplinary approach to long-term modelling for power system expansion. *International Journal of Critical Infrastructures, 3*, 235–264.

Dimitrovski, A., Tomsovic, K., & Ford, A. (2007b). *Comprehensive long term modeling of the dynamics of investment and network planning in electric power systems* (pp. 235–264)

Dyner, I. (2001). From planning to strategy in the electricity industry. *Energy Policy*, 1145–1154

Dyner, I., & Bunn, D. W. (1997). A system simulation platform to support energy policy in Columbia. In Systems Modelling (Ed.), *for*. Energy Policy, Chichester: Wiley.

Dyner, I., Larsen, E. R., & Franco, C. J. (2009). *Games for electricity traders: Understanding risk in a deregulated industry. Energy Policy* (pp. 465–471).

Dyner, I., Smith, R., & Pena, E. (1995). System dynamics modeling for residential energy efficiency and management. *Journal of the Operational Research Society, 46*, 1163–1173.

Feng, Y., Chen, S. Q., & Zhang, L. X. (2013). System dynamics modeling for urban energy consumption and CO_2 emissions: A case study of Beijing-China. *Ecological Modelling, 252*, 44–52.

Fiddaman, S. (2002). Exploring policy options with a behavioral climate-economy model. *System Dynamics Review, 18*(2), 243–267.

Ford, A. (1996). System dynamics and the electric power industry. *System Dynamics Review, 13* (1), 57–85.

Ford, A. (2001). Waiting for the boom: A simulation study of power plant construction in California. *Energy Policy, 29*(11), 847–869.

Ford, A., Vogstad, K., & Hilary, F. (2007). Simulating price patterns for tradable green certificates to promote electricity generation from wind. *Energy Policy, 35*, 91–111.

Forrester, J. (1961). *Industrial dynamics*. Cambridge, USA: MIT Press.

Han, J., & Hayashi, Y. (2008). A system dynamics model of CO_2 mitigation in China's inter-city passenger transport. *Transportation Research Part D: Transport and Environment, 13*, 298–305.

Hasani, M., & Hosseini, S. H. (2011). Dynamic assessment of capacity investment in electricity market considering complementary capacity mechanisms. *Energy, 36*(1), 277–293.

Jin, W., Xu, L., & Yang, Z. (2009). Modeling a policy making framework for urban sustainability: Incorporating system dynamics into the ecological footprint. *Ecological Economics, 68*(12), 2938–2949.

Naill, R. (1973). The discovery life cycle of a finite resource: A case study of U.S. natural gas, in toward global equilibrium. In D. Meadows & D. Meadows (Eds.). Waltham, MA: Pegasus Communications.

Ochoa, P. (2007). Policy changes in the Swiss electricity market: A system dynamics analysis of likely market responses. *Socio-Economic Planning Sciences, 41*(4), 336–349.

Olsina, F., Garces, F., & Haubrich, H.-J. (2006). Modeling long-term dynamics of electricity markets. *Energy Policy, 34*(12), 1411–1433.

Park, J., Ahn, N.-S., Yoon, Y.-B., Koh, K.-H., & Bunn, D. W. (2007). Investment incentives in the Korean electricity market. *Energy Policy, 35*(11), 5819–5828.

Pasaoglu, G., & Or, I. (2006). A system dynamics model for the decentralized electricity market. *International Journal of Simulation Systems Science and Technology, 7*(7), 40–55.

Qudrat-Ullah, H. (2005). MDESRAP: A model for understanding the dynamics of electricity supply, resources and pollution. *International Journal of Global Energy, 23*(1), 1–13.

Qudrat-Ullah, H. (2013). Understanding the dynamics of electricity generation capacity in Canada: A system dynamics approach. *Energy, 59*, 285–294.

Qudrat-Ullah, H. (2014). *Better decision making in complex, dynamics tasks*. USA, New York: Springer.

Qudrat-Ullah, H. (2015). Independent power (or pollution) producers? electricity reforms and IPPs in pakistan.*Energy, 83*(1), 240–251.

Qudrat-Ullah, H., & Davidsen, P. (2001). Understanding the dynamics of electricity supply, resources and pollution: Pakistan's case. *Energy, 26*(6), 595–606.

Saeed, K. (2013). Managing the energy basket in the face of limits: A search for operational means to sustain energy supply and contain its environmental impact. In H. Qudrat-Ullah (Ed.), *Energy policy modeling in the 21st century* (pp. 69–86). NYm USA: Springer.

Sterman, J. (2000). *Business dynamics: Systems thinking and modeling for a complex world*. NY, USA: McGraw-Hill.

Trappey, A., Trappey, C. V., Lina, G. Y. P., & Chang, Y. S. (2012). The analysis of renewable energy policies for the Taiwan Penghu island administrative region. *Renewable and Sustainable Energy Reviews, 16*, 958–965.

van Ackere, A., & Ochoa, P. (2009). Policy changes and the dynamics of capacity expansion in the Swiss electricity market. *Energy Policy, 37*(5), 1983–1998.

Vennix, J. (1996). *Group model building: Facilitating team learning using system*. Chichester, England: Wiley.

Vlahos, K. (1998). The electricy markests microworld. Versión 1.0. LBS, UK.

Vogstad, K., Botterud, A., Maribu, K. M., & Jensen, S. G. (2002). The transition from fossil fuelled to a renewable power supply in a deregulated electricity market. *Science And Technology*.

Chapter 4
Understanding the Physics of Stocks and Flows

I have found out what economics is; it is the science of confusing stocks with flows

—Michael Kalecki

Human decision making and energy policy making have mostly been about managing various stocks of often limited resources (e.g., capital, people, money, technology). These stocks can only be managed by controlling their flows (i.e., decision makers can influence these stocks by regulating flows). These stocks and flows over time give rise to various dynamics of an energy system (i.e., imbalance between demand and supply of electricity increases (or decreases), renewable electricity generation capacity increases but with an unexpected rise in electricity price). Therefore, we begin to unfold the physics of underlying stocks and flows of various feedback loops of energy systems.

4.1 Physics of a Positive Feedback Loop: A Growth-Propelling Behavior

Consider a generic positive feedback loop, **R**, (as shown in Fig. 4.1), consisting of a single stock, Level_Any Stock, a single inflow, Rate_ Any Flow, and a constant, Any_Constant. The physics of this basic positive feedback loop, the simplest feedback system, is represented by the corresponding stock and flow diagram[1] as shown in Fig. 4.1. Here, we have only one stock, denoted by L. This stock accumulates its net flow rate (in this case, it is net inflow). The net inflow rate depends on the state of L. Mathematically:

$$L_t = \text{Stock at any time,} \quad t = L_0 + \int_0^t (\text{Net Inflow}) \, dt,$$

---

[1]The "cloud" appearing in this stock and flow diagram could mean that we are not modeling the source of flow (i.e., Rate_Any Flow).

© The Author(s) 2016
H. Qudrat-Ullah, *The Physics of Stocks and Flows of Energy Systems*,
SpringerBriefs in Complexity, DOI 10.1007/978-3-319-24829-5_4

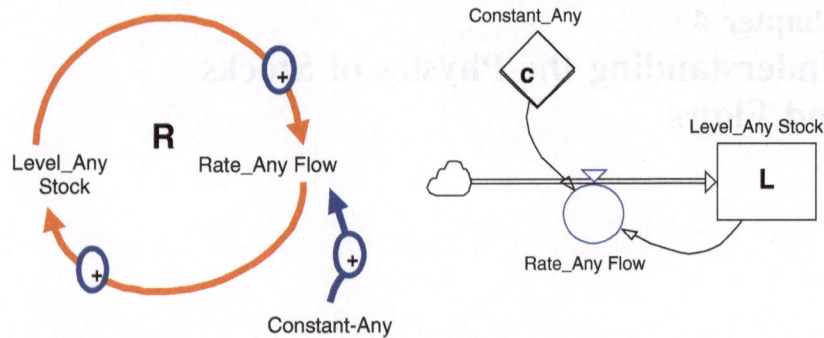

Fig. 4.1 The representation of the physics of a positive feedback loop

where L_0 is the value of initial stock.

It is clear that net inflow is proportional to the size of the stocks. We have:

$$\text{Net Inflow} = f(L)$$

Now for a linear system, net flow is directly proportional to the state of the stock:

$$dL/dt = c * L,$$

where the constant c represents the fractional growth rate of the stock.

Using integral calculus, we can find the solution of this differential equation as

$$L_t = L_0 * e^{(ct)}$$

This solution indicates that the stock will grow at an exponential growth rate (as shown in Fig. 4.2). That is why we call a positive feedback loop, a growth-propelling feedback system. Although we are able to find an analytical

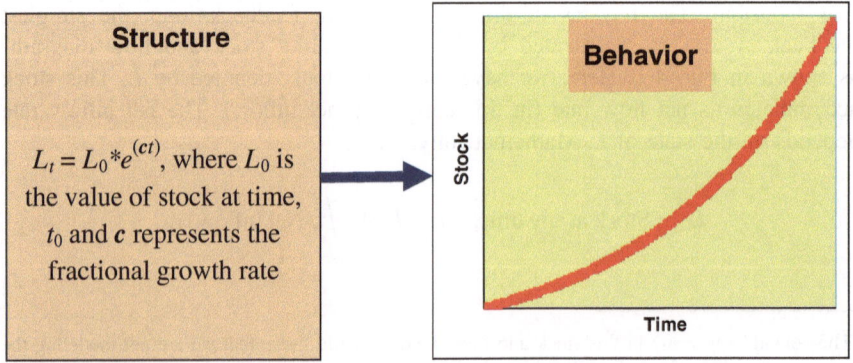

Fig. 4.2 Structure-behavior graph of a basic positive feedback loop

Table 4.1 Powersim equations for the "Debt Accumulation Process"

Name	Units	Definitions/Equations
Level_Any Stock	Million$	40
Rate_Any Flow	Million$/year	'Level_Any Stock' * Constant_Any
Constant_Any	1/year	0.05

solution to such simple feedback systems, this possibility becomes almost impossible when the number of stocks and flows increases (and it does so in the case of many real work systems including energy systems) or the relationship between variables becomes more complex (than the simple linear as we considered in our illustration). Therefore, simulation methods are used to understand the dynamics of complex systems better.

Now, consider the example of the "debt accumulation process", where a company borrows money at 5 % with an initial amount of US$40 million. The behavior of this debt accumulation process is shown in Fig. 4.2. Table 4.1 lists the set of equations for this basic feedback system that are written in Powersim™, a simulation platform for system dynamics models.

4.2 Physics of a Negative Feedback Loop: A Goal-Seeking Behavior

Whereas a positive feedback loop seeks an accelerating growth, a negative feedback loop gives rise to a goal-seeking behavior in which the state of a stock variable is driven to a particular value. Consider a generic negative feedback loop, **B** (as shown in Fig. 4.3), consisting of a single stock, Level_Any Stock, L, a single outflow, Rate_ Any Flow, and a constant, Any_Constant, d. The physics of this basic negative feedback loop is represented by the corresponding stock and flow diagram

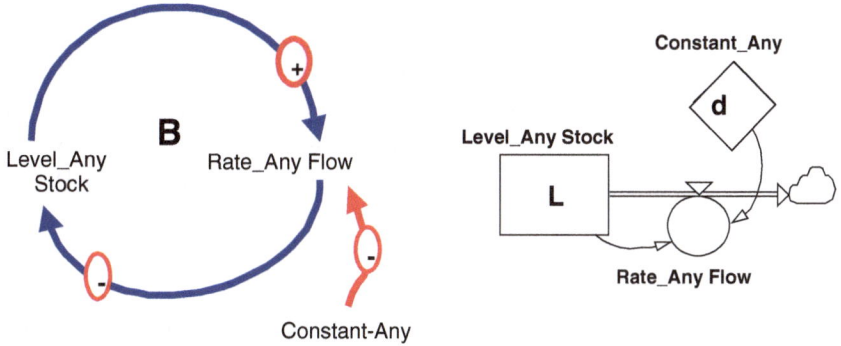

Fig. 4.3 The representation of the physics of a negative feedback loop

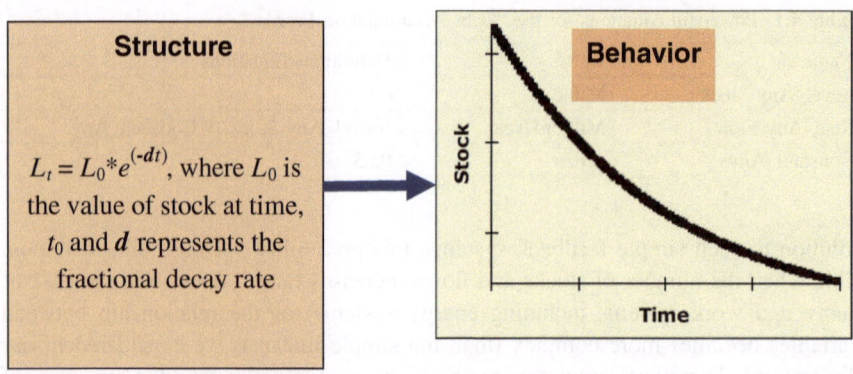

Fig. 4.4 Structure-behavior graph of a basic negative feedback loop

as shown in Fig. 4.3. Here, we have only one stock, denoted by L. The state of this stock changes through its net inflow rate. Now for a linear system, net outflow is directly proportional to the state of the stock:

$$\text{Net Inflow} = dL/dt = -\text{Net Outflow} = -d * L$$

where the constant, d, represents the fractional decay rate of the stock, L.

The solution of this differential equation is:

$$L_t = L_0 * e^{(-dt)}$$

This solution indicates that the stock will decay at an exponential rate (as shown in Fig. 4.4). That is why we call a negative feedback loop, a goal-seeking feedback system. In the absence of any stated goal, stock will decay to zero (as in this case).

Now, consider the example of a "capital depreciation process", where initial production capital is 100 MW and it depreciates at 5 % per year. The behavior of this capital depreciation process is shown in Fig. 4.4. Table 4.2 lists the set of equations for this basic feedback system that are written in Powersim.[2]

There are various stocks in an energy system that decision makers want to keep at a specific level. For instance, fuel inventory needs to be maintained at a specific level to avoid extra storage cost or production loss due to the shortage of fuel. Therefore, we need to understand the dynamics of a negative feedback loop with the stock, L, that has to be maintained at a specific level, $L*$ (as shown in Fig. 4.5). It takes time, Adjustment Time, t, to close this gap. The physics of this basic

---

[2]Powersim[TM] is a trademark of Powersim Software Inc. (http://www.powersim.com). It provides a platform to create stock and flow based simulation models. There are several others who provide the similar tools.

Table 4.2 Powersim equations for the "Capital Depreciation Process"

Name	Units	Definitions/Equations
Level_Any Stock	MW	100
Rate_Any Flow	MW/year	'Level_Any Stock' * Constant_Any
Constant_Any	1/year	0.05

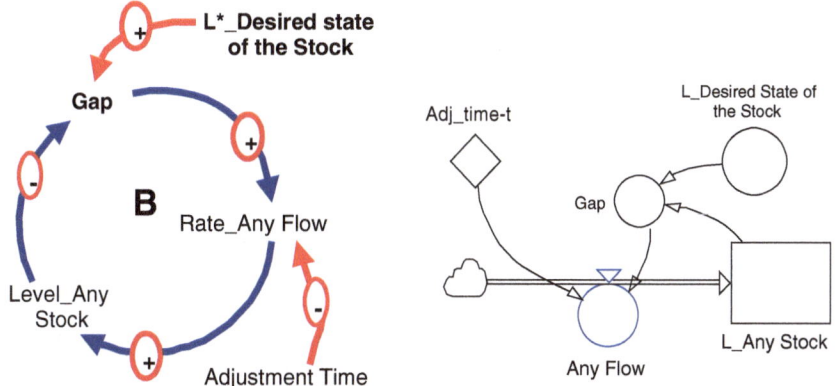

Fig. 4.5 Diagramming the physics of a negative feedback loop

negative feedback loop is represented by the corresponding stock and flow diagram as shown in Fig. 4.5. Although the net inflow to the stock will vary according to the size of the gap remaining, following Sterman's pioneering work (Sterman 2000), we assume it to be a constant equal to the average gap:

$$\text{Net Inflow} = \text{Gap}/t = (L* -L)/t$$

Depending on size of the existing stock and desired stock, there is a gap that a negative feedback loop attempts to close. There can be two situations: (i) the existing state of the stock, L, is lower than the desired state, $L*$ or (ii) the existing state of the stock, L, is higher than the desired state, $L*$. In both cases, the state of the stock is driven exponentially to the desired state of the stock. Consider the case of an oil-fired power plant company that needs to maintain a fuel inventory of 20 million tons. Figure 4.6 shows two exponential trajectories of the state of the stock to reach the target level, 20 million tons: one from a higher, 35 million tons, initial state of the stock, and the other from a lower, 5 million tons, initial state of the stock.

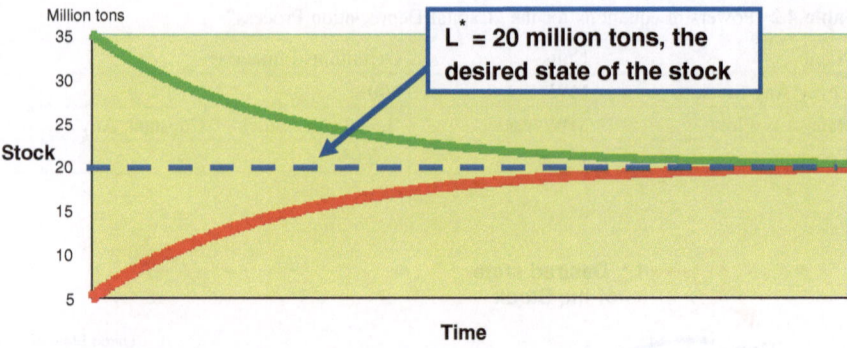

Fig. 4.6 Dynamics of a basic balancing feedback loop system

4.3 Dynamics of Material Delays and Information Averaging Processes

Delays are pervasive in energy systems. For instance, it takes time to get approval for the site, to build a power plant, to commission it, to train plant personnel, and to build a waste disposal facility. These delays (or time lags) come in two forms: material delay (e.g., physical materials have to move from the source, e.g., a supplier's site, to the destination, e.g., plant site); and information delay (e.g., it takes time get approval of site, financing, and commissioning permit). Delays like stocks decouple various inflows and outflows of any energy systems. In the following, we describe the structure and behavior of these delays so we may better understand the dynamics of various feedback loops of energy systems.

4.3.1 Basic Dynamics of Material Delays in Energy Systems

Intuitively, it is clear that the structure of a delay must have a stock (where material in transit is kept). For example, when IPPs invest in large power plants the average delay between the time the construction of a power plant begins and the time when this power plant becomes operational is 6 years and the average life of a gas-powered plant is 40 years (Qudrat-Ullah 2015).

It must be noted that in a material delay the law of conservation of mass holds; that is, all the output from the delay stock is equal to the input going into that delay stock. In other words, nothing is created nor is destroyed while in transit from one stock to the other. Consider a simple example of two stocks of gas-based power plants: capital stock under construction (KUC), and capital stock, K, ready to produce electricity.

KUC increases by the new investments (KI_r) made by the IPPs and decreases when the construction of the power plants is completed. The construction time, CT,

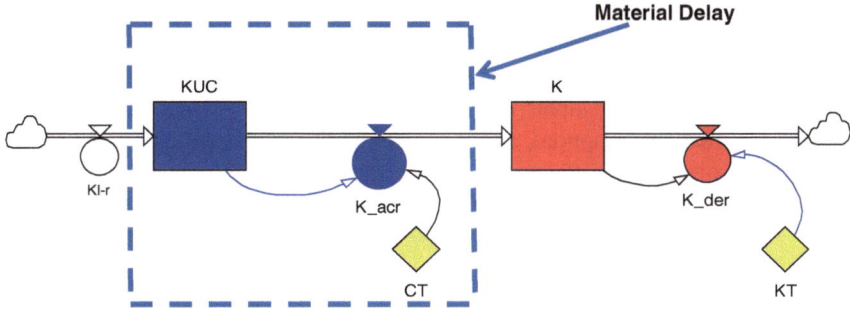

Fig. 4.7 Stock and flow diagram of a material delay

is 6 years in this example. Please note that capital investments (KI) are made to account for both the new demand and the depreciated production capital. As soon as the power plants become ready they add to the capital (production) stock. Thus, K increases by the capital acquisition rate (K_{acr}), an inflow to K, and decreases by the capital depreciation rate (K_{der}), an outflow to K. The average life of power plant, KT, is 40 years. Figure 4.7 represents a stock and flow diagram of this material delay.

Figure 4.8 displays the specific dynamics of construction delays: 6 years versus 12 years. In this example, we assumed that the initial amount of the capital under construction is 5000 MW electricity generation capacity (i.e., KUC = 5000 MW) and no new investments are made (i.e., KI_r = 0 MW/year). To understand the dynamics of material delay better (i.e., delayed movement of material from one stock, KUC, to the other, K), we have also switched off the outflow from K. Then, depending upon how big or small the outflow, K_{acr} is, all the material in KUC is shipped or emptied into the electricity production capital, K. When CT is 6 years old, KUC empties at a faster rate as compared with the case when construction time is assumed as 12 years. Likewise, the production capital, K, builds up from zero

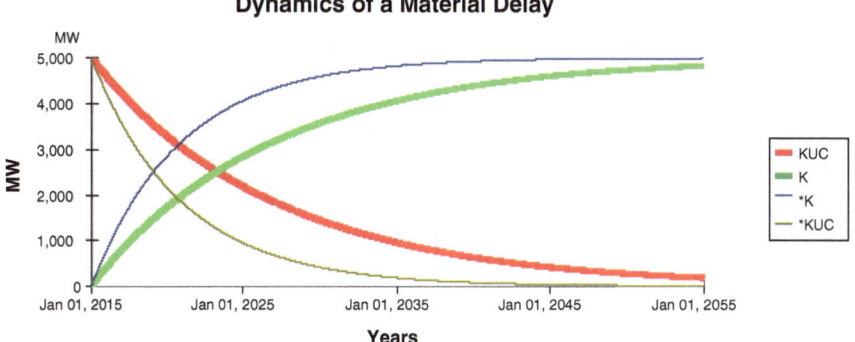

Fig. 4.8 Dynamics of construction delays

MW to 5000 MW capacity level at a faster (when CT = 6 years) or slower rate (when CT = 12 years). Thus, we can see that (i) material delays create time lags in changing the level of corresponding stocks, and (ii) material delays behave as stocks (i.e., the material inputs are stored first, while in transit, before they are released for final consumption), and (iii) material delays strictly follow the law of conservation of mass.

The mathematical model of this delay process can be written as:

$$(\partial/\mathrm{d}t)\mathrm{KUC} = -K_{\mathrm{acr}},$$
$$(\partial/\mathrm{d}t)K = K_{\mathrm{acr}} - K_{\mathrm{der}}$$
$$K_{\mathrm{acr}} = \mathrm{KUC}/\mathrm{CT},$$
$$K_{\mathrm{der}} = K/KT,$$
$$\mathrm{HUC}_{t=0} = 5000 \text{ MW and } K_{t=0} = 0.$$

It is worth mentioning that we have demonstrated the dynamics of simple first-order (i.e., it involved only one stock or mathematically speaking, it has a first-order differential equation representation) material delay. However, delays can be of higher order and with different dynamics associated with them. The interested reader should consult the excellent treatment on delays by Richardson and Pugh III (1981).

4.3.2   Structure and Behavior of Information Delays of Energy Systems

In contrast to a material delay, the transmission of information is delayed but not conserved. In fact, when the duration of a delay increases even substantially, the material is conserved but, to the contrary, long delayed information often gets forgotten. Therefore, we need a different structure to model an information delay.

Consider the situation where you want to acquire large capital equipment (e.g., a nuclear power plant). Certainly, it would be a time-consuming activity. For instance, it takes time to get regulatory approvals, shipment license, financing approvals, and training of the required workforce. As a manager of a large utility firm, you don't want to commit resources to various sectors (e.g., workforce hiring and training) without having better clarity around the various approval processes. To deal with the uncertainty and irregularities of these processes, the decision makers often resort to the averaging process. A common averaging process that gives higher weight to the recent data is called "exponential smoothing" or "adaptive expectation" (Sterman 2000). Notice that our expectations rarely change instantaneously. Rather, once new information becomes available, our beliefs get updated. Thus, the belief or perception adjustment is a gradual process and the

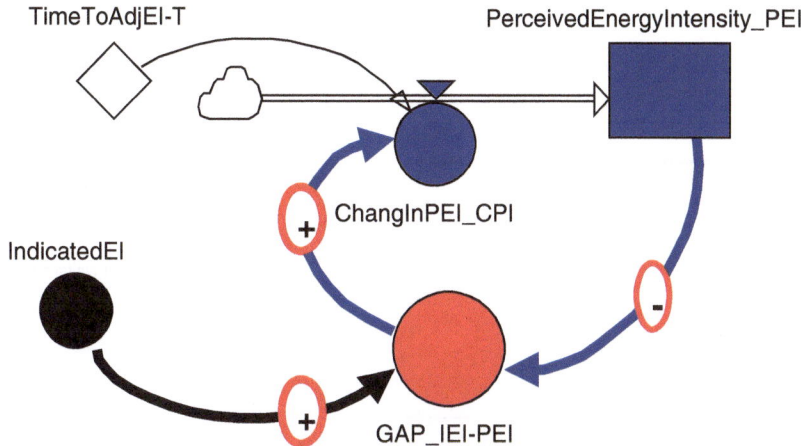

Fig. 4.9 Feedback structure of an information delay

perceived value of input is a stock. Consider the example of perceived energy intensity (PEI) as shown in Fig. 4.9.[3]

In this feedback structure of adaptive expectations, the PEI is a stock and the rate of change in the perception of energy intensity (CPI_r) is proportional to the gap between the indicated or reported value of energy intensity IEI, and the PEI with T as time it takes for the adjustment in the perceptions or expectations. Mathematically, we can write the process of this information delay as:

$$\text{PEI} = \int [(\text{IEI} - \text{PEI})/T]\mathrm{d}t + \text{PEI}(t = 0)$$

In fact, this feedback structure of information delay represents a negative feedback loop system that we have discussed in Sect. 4.2. Here, in Fig. 4.10, we can see the behavior of this information delay. There are two scenarios, one with the reported energy intensity (IndicatedEI) equal to 0.20 MWh/$ and the other showing a step increase in the reported energy intensity where it is equal to 0.30 MWh/$. In both cases, we can see that only after a considerable time lag, the perceptions (i.e., the perceived energy intensity) are adjusted to the reported values, a typical behavior of a negative feedback loop.

---

[3]This diagram, which combines both a "stock and flow diagram" and a "casual loop diagram", is called a hybrid diagram. In these diagrams, one can not only see the casual feedback structure (e.g., here we see a negative feedback loop) as with any casual loop diagram will do, but also with an explicit representation of stocks and flows.

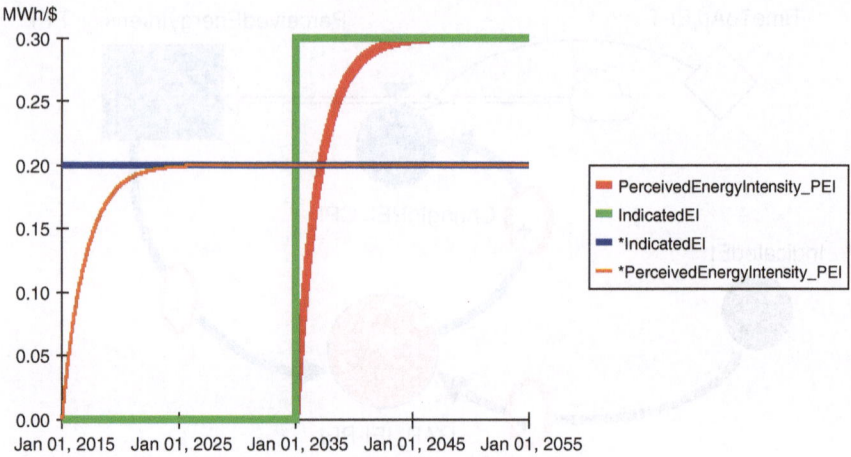

Fig. 4.10 Dynamics of a simple first-order information delay

4.4   Dynamics of Nonlinearities of Energy Systems

A process is said to be a nonlinear process if the process outcome is not proportional to the process input. In the context of energy systems planning and decision making, there are several nonlinear processes that need to be carefully understood. For instance, incentives for renewable energy (e.g., feed-in tariffs) are often graduated. That is, when the electricity produced from renewables doubles, the incentive payment rate increases but never exactly doubles. Likewise, when the demand of energy increases, the additional capacity is added to the production systems, maybe proportionally. However, once the new capacity is fully operational, it sets the limit on how much (maximum) electricity could be produced regardless of the increase in the demand for electricity, a nonlinear response.

In system dynamics models, a table function is used to incorporate such nonlinear processes, where the relationship between the two variables, the independent and dependent, is specified as a table of values. For instance, Fig. 4.11 displays the nonlinear response of a variable, Import Dependency Premium,[4] in response to the input variable, Import Dependency and Table 4.3[5] specifies the values of these two variables. When Import Dependency increases from 0 to 16 %, the response variable, Import Dependency Premium, increases at a much faster rate (i.e., an exponential growth rate). Then, when the import dependency increases from 16 to 82 % (approximately), the premium increases almost linearly, exhibiting a linear

---

[4]In our earlier wok (Qudrat-Ullah 2005), we identified this premium as implying that there is a preference for the least import-dependent technology for electricity generation technology.

[5]Different system dynamics modeling platforms (e.g., Powersim™, Vensim™, iThink™, Stella™) provide ways to incorporate this table function in a stock and flow based simulation model.

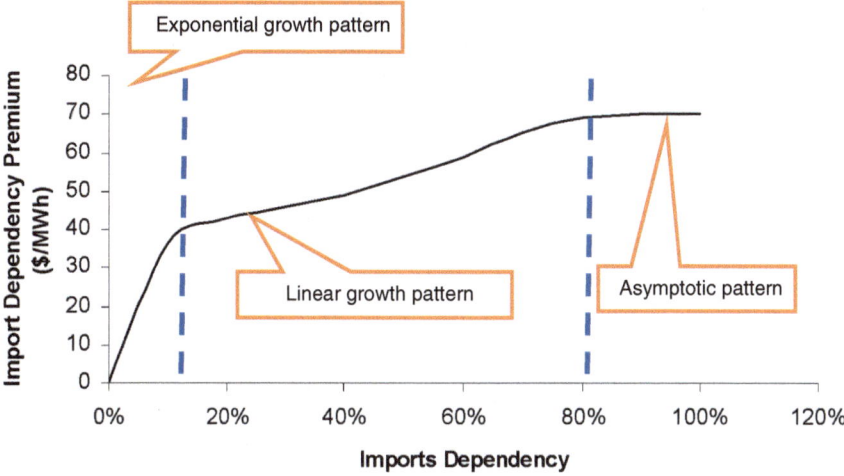

Fig. 4.11 Dynamics of a nonlinear relationship between import dependency premium and imports dependency. *Source* Adopted from Qudrat-Ullah (2005)

Table 4.3 Specified values for the import dependency and the premium

Variables	Values
Dependency (%)	(0, 10, 20, 30, 40, 50, 60, 70, 80, 90, 100)
Dependency premium ($/MWh)	(0, 30, 40, 45, 50, 55, 60, 65, 70, 70, 70)

growth pattern. Finally, we can see the asymptotic behavior of the premium, again a nonlinear response. Thus, the structure of a nonlinearity (e.g., as given in Table 4.3) is responsible for a nonlinear behavior (e.g., as shown in Fig. 4.11).

4.5 Summary

To understand the physics of the stocks and flows of energy systems, it is critical to develop an appreciation and a thorough understanding of the underlying structures including various feedback loops, time delays, uncertainties, and nonlinearities. Essentially, these structures are responsible for the dynamic behavior of energy systems that the energy policy makers are interested to monitor, control, and manage. If the relationship between any two variables of energy systems are linear than we have great mathematical models and tools to understand their behavior. However, most of the energy policy related issues and tasks are highly complex, nonlinear, and dynamic. To understand their behavior we need to identify the nature of the feedback structures, create mathematical models (i.e., equations to represent various stocks and flows, time delays, and nonlinear functions), and simulate them.

Therefore, an understanding of the physics of these underlying structures is hoped to help the energy planners and decision makers to manage the available, but limited energy resources better. To further elaborate on the physics of stocks and flows, in the next chapter we present the modeling process for some key energy system processes.

References

Qudrat-Ullah, H. (2005). MDESRAP: A model for understanding the dynamics of electricity supply, resources and pollution. *International Journal of Global Energy Issues, 23*(1), 1–13.

Qudrat-Ullah, H. (2015). Independent power (or pollution) producers? *Electricity Reforms and IPPs in Pakistan, Energy, 83*(1), 240–251.

Richardson, G. P., & Pugh II., A. B. (1981). Introduction to system dynamics modeling with DYNAMO. Cambridge, MA: The MIT Press. Reprinted by Pegasus Communications, Cambridge, MA: USA.

Sterman, J. (2000). *Business dynamics: Systems thinking and modeling for a complex world.* NY, USA: McGraw-Hill.

Chapter 5
On the Modeling of Key Structural Process of Energy Systems

Everything must be taken into account. If the fact will not fit the theory—let the theory go

—Agatha Cristie

Armed with the fundamental understanding of feedback loop structures from the material of Chap. 4, now we turn our attention to models and explain the dynamics of various accumulation and structural processes (i.e., stocks, flows, time delays, and nonlinearities) that are common to most of the energy systems planning and decision-making situations. We achieve such understanding with the help of stock and flow representations, structural representations and mathematical formulations, and behavioral outputs of various fundamental processes of energy systems including energy demand, policy incentives and interfuel substitution mechanisms, manpower recruitment and training, energy production, and environmental emissions and CO_2 tax.

5.1 Stocks and Flows of Energy Demand Process

There are various ways that the modeling community has incorporated the energy demand process in their models. However, when it comes to energy policy modeling, system dynamics models have modeled the energy demand process often endogenously. A common feedback structure with endogenous energy demand is shown in Fig. 5.1.

In the feedback structure of energy demand shown in Fig. 5.1, the GDP (gross domestic product of a nation) is assumed to be exogenous. Often, the GDP's historical trend is modeled in system dynamics and other modeling methodologies. In this model of energy demand, the average energy intensity (EI_{avg}) is captured as an exponential smoothing of the short-term indicated energy intensity (EI_{sti}) over a period (T_{stip}). This exponential smoothing process is an example of an information averaging process, that is, an information delay. Mathematically, it can be written as

© The Author(s) 2016
H. Qudrat-Ullah, *The Physics of Stocks and Flows of Energy Systems*,
SpringerBriefs in Complexity, DOI 10.1007/978-3-319-24829-5_5

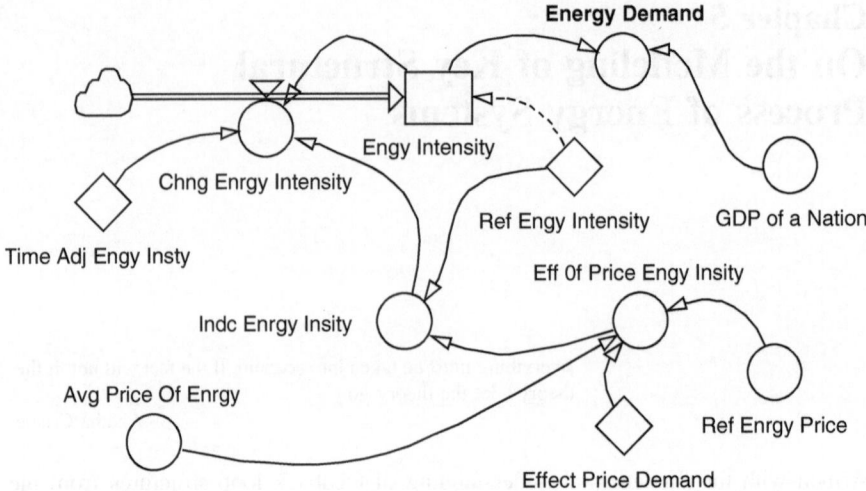

Fig. 5.1 Stock and flow diagram of energy demand process

$$\text{Energy Demand} = \text{EI}_{\text{avg}} * \text{GDP}$$
$$(d/dt)\text{EI}_{\text{avg}} = (\text{EI}_{\text{sti}} \cdot \text{EI}_{\text{avg}})/T_{\text{stip}},$$
$$\text{EI}_{\text{avg}}(t_0) = \text{EI}_{\text{sti}}(t_0)$$

The reference energy intensity and the effect of energy price on electricity intensity determine the indicated energy intensity. The reference energy intensity represents the case-specific base year value. The effect of energy price on energy intensity is dependent upon three factors:

- The effect of energy price on demand
- The average energy price
- The reference electricity price.

If the price of energy decreases (relative to the reference energy price), the energy intensity exhibits a growth pattern. Conversely, energy consumption declines when the price rises. Contrary to this endogenous approach for modeling energy demand, some researchers have treated energy demand as an exogenous input to their dynamic model (e.g., please see Ford 1996 and Saeed 2013). Depending upon the purpose and focus of the dynamic model, one could decide on the appropriate modeling approach for energy demand.

As an example, Fig. 5.2 displays the dynamics of the key accumulation process, a stock, Electricity Intensity. This example is based on our earlier work in this stream (Qudrat-Ullah 2014). Here, the reference electricity intensity represents our case-specific base year (2000) value. The year 2000 was chosen because Ontario's

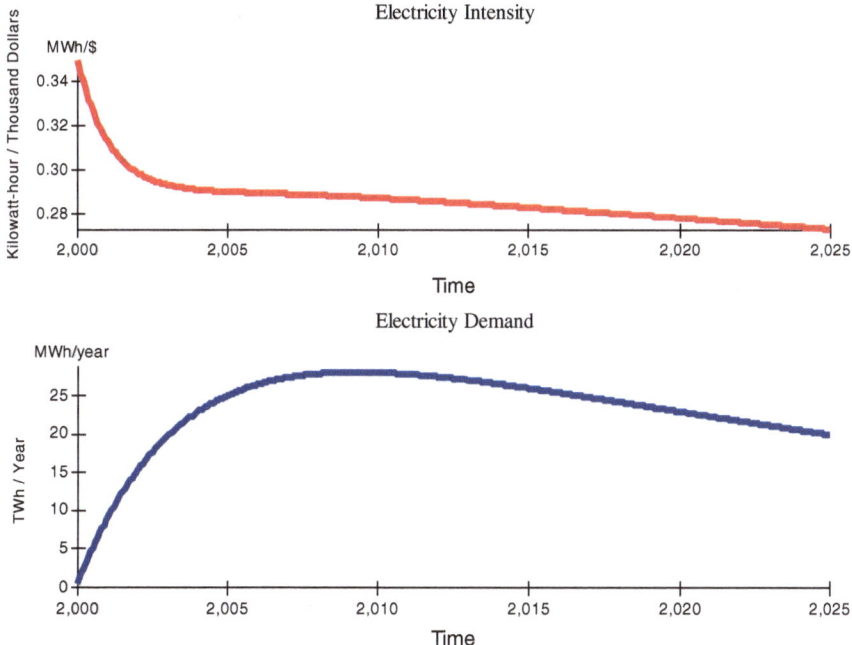

Fig. 5.2 Dynamics of electricity intensity and electricity demand

Environmental Labeling Program began in 2000 (OEB 2002). The effect of electricity price on electricity intensity depends on (i) the effect of electricity price on demand, (ii) the average electricity price, and (iii) the reference electricity price and is modeled as

$$\text{Effect of Electricity Price on EI} = e^{\,(\text{Effect Of Elec Price On Demand})\,*\,\text{Ln(Avg Price Of Elec/Ref Elec Price)}}$$

The effect of price on demand is a parameter whose value is based on long-term Ontario experience and is estimated by our model as "0.41" (we input the reference data in the model and estimated this parameter. With this value, our model was able to reproduce both electricity demand and price data for Ontario for the period 2000–2012). The reference electricity price is an extrapolation of actual prices from 2000 to 2012, and the average electricity price is computed endogenously in the model. If the price of electricity decreases relative to the reference electricity price, the electricity intensity exhibits a growth pattern. Conversely, the electricity consumption declines when the price rises.

5.2   Modeling of Interfuel Substitution Process and Policy Incentives

The interfuel substitution mechanism whereby a fuel source is substituted by another competitive fuel (e.g., substitution of fossil fuels by renewables) is an issue of substantial interest to the energy policy decision makers (Qudrat-Ullah 2005; Abada et al. 2013). In general, IPPs are interested in investing in those electricity-generating technologies that provide them with a quick and higher rate of return. In their decisions to invest in a particular technology, the policy incentives offered by the host agency or organization or nation play a key role. Moxnes (1990) introduced a structure that includes various cost elements related to each competing technology and assigns the respective share of investment to each technology with preference for the least-cost technology. For instance, consider the case where, in addition to the standard cost elements, the investment incentive premium (IIP), the import dependency premium (IDP), and the environment premium (ENP) for each technology are also to be considered. The standard cost elements include capital costs (CC), operating costs (OC), and fuel costs (FC). Once the total cost for each technology (TC) is determined, we apply a multinomial logit (MNL) model to obtain the share for each technology (ST). Building on Moxnes (1990) model, we can have a mathematical model as

$$TC_i = CC_i + OC_i + FC_i + IIP_i + IDP_i + ENP_i,$$
$$ST_i = \exp(-\alpha \times TC_i)/\Sigma\exp(-\alpha \times TC_i),$$

where i = technology 1, technology 2, technology 3,... technology n.

The MNL has only one parameter α (distribution parameter). When α takes an extreme value, then the aggregate choice of the whole sector (all the IPPs) for a technology mimics the choice of an individual IPP. The lower values of α represent greater variation among the individual choices and the aggregate choice. When the total costs of all the technologies are equal, the market share is split into equal shares.

To understand the dynamics of the interfuel substitution process, we present a case from our earlier work (Qudrat-Ullah 2005), where in addition to the standard cost elements, an environment premium incentive is modeled. Four technologies, coal, hydro, oil, and gas-based power plants are competing against each other in this substitution process. Figure 5.3 displays the feedback structure responsible for generating the share of each technology (Fig. 5.5) based on costs and policy incentives.

As the investments are made for electricity production capital, after the related construction and approval delays, new production capital becomes available. In system dynamics modeling, it is common to use a two-vintage structure to portray the aging process of installed capacity (Moxnes 1990; Qudrat-Ullah 2013). Figure 5.4 presents this aging process in a stock and flow diagram. The power producers' investment in each of the electricity-generating capitals (EC) are

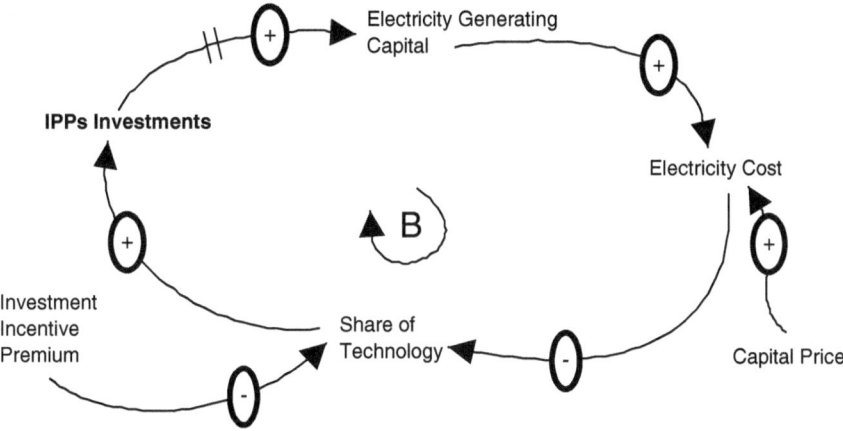

Fig. 5.3 Share of investments for each technology due to policy incentives

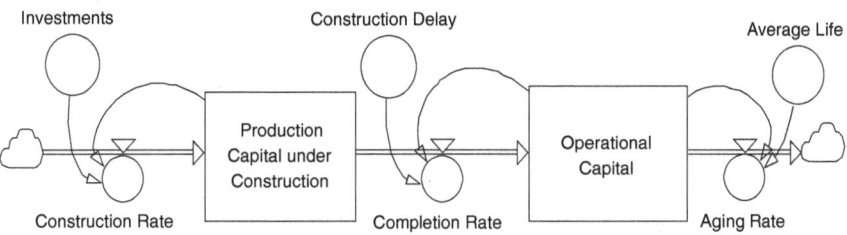

Fig. 5.4 A two-vintage structure of production capital. *Source* Qudrat-Ullah (2013)

represented as an adjustment of indicated investments in electricity-generating capital (IEC) over the adjustment period (Tad_j). The indicated investments are determined by the total demand of electricity-generating capital (TEC), unit capital cost (UCEC), and the share (SEC) of each of the competing technologies. Mathematically, the production capital accumulation process is represented as

$$d(EC_i)/dt = \big((IEC_i - EC_i)/Tad_j\big), \quad EC_i(t_0) = IEC_i(t_0),$$

where i = thermal, hydro, others, nuclear.

Here, due to the imposition of CO_2 control, the future investments are made in the hydro power plants (Table 5.1; Fig. 5.5). The cost of electricity is relatively increased (due to CO_2 tax and the capital-intensive technology substitution); relatively less generation capacity (hydro power plants) is added after the construction delay that eventually causes the loss in the demand for electricity.

Table 5.1 and Fig. 5.5 show a drastic shift in the capacity mix. There is a decrease of 59, 63, and 69 % in the cumulative capacity of coal-based, oil-based, and gas-based power plants, respectively. On the other hand, the hydro-based

Table 5.1 Substitution process (reference scenario vs. environment scenario)

Scenario	Coal-based	Hydro-based	Oil-based	Gas-based
Reference scenario (GW)	0.0644	18.34	0.755	25.18
Environment-oriented scenario (GW)	0.0263	33.12	0.278	7.83
Percent change	−59	+81	−63	−69

Source Qudrat-Ullah (2005)

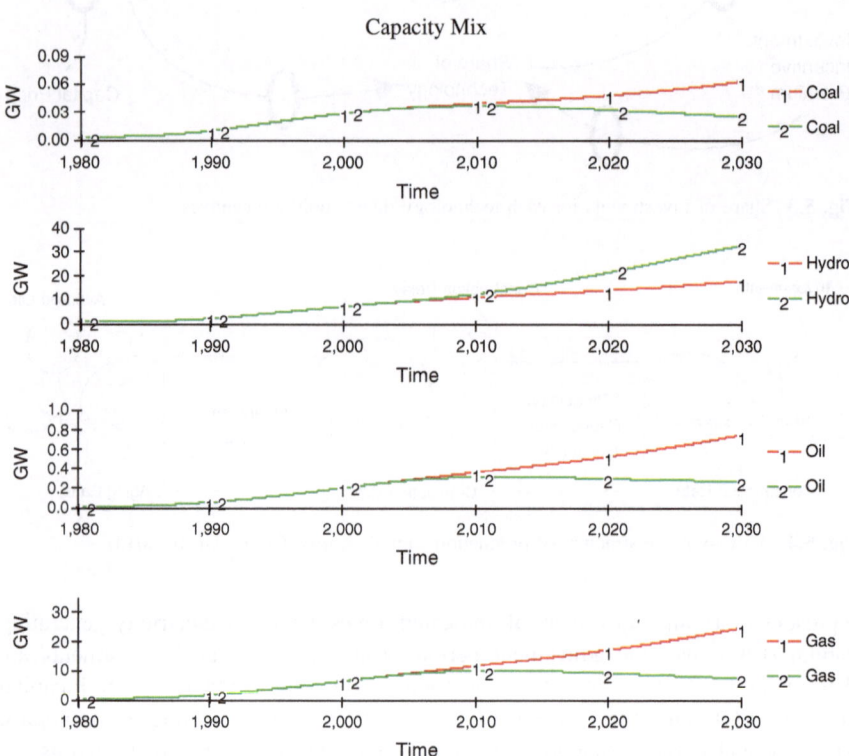

Fig. 5.5 Dynamics of the substitution process: reference (*1*) versus environment (*2*) scenario. *Source* Qudrat-Ullah (2005)

production capacity experiences a significant increase of 81 %, so as to gain a market share of 80 %.

Although our model, as elaborated above, of the substitution process and policy incentive included only four technologies, the procedure is general enough to include any number of electricity-generating technologies in the model. Therefore, by understanding these long-term dynamics of interfuel substitution process and various policy incentives, both the energy policy evaluators and designers can improve their decision making regarding the management of limited energy resources: the crux of any effective energy policy.

5.3   Stocks and Flows of the Manpower Recruitment and Training Processes

Much like any industry, trained manpower planning plays a fundamental role in the successful operations of any type of power plant (e.g., coal, oil, hydro, nuclear, and the alternate fuel-based power plants). Both the timely recruitment and the effective training of newly hired personnel are the two key business processes for any organization. Both of these processes involve time lags: it takes time to hire and train people.

To understand better the physics of the stocks and flows involved in manpower planning of a power plant, we need to identify the underlying feedback structures. Figure 5.6 shows a two-level manpower structure: Level 1 (L1) represents the stock of newly hired personnel, and Level 2 (L2) is the stock (pool) of trained people.

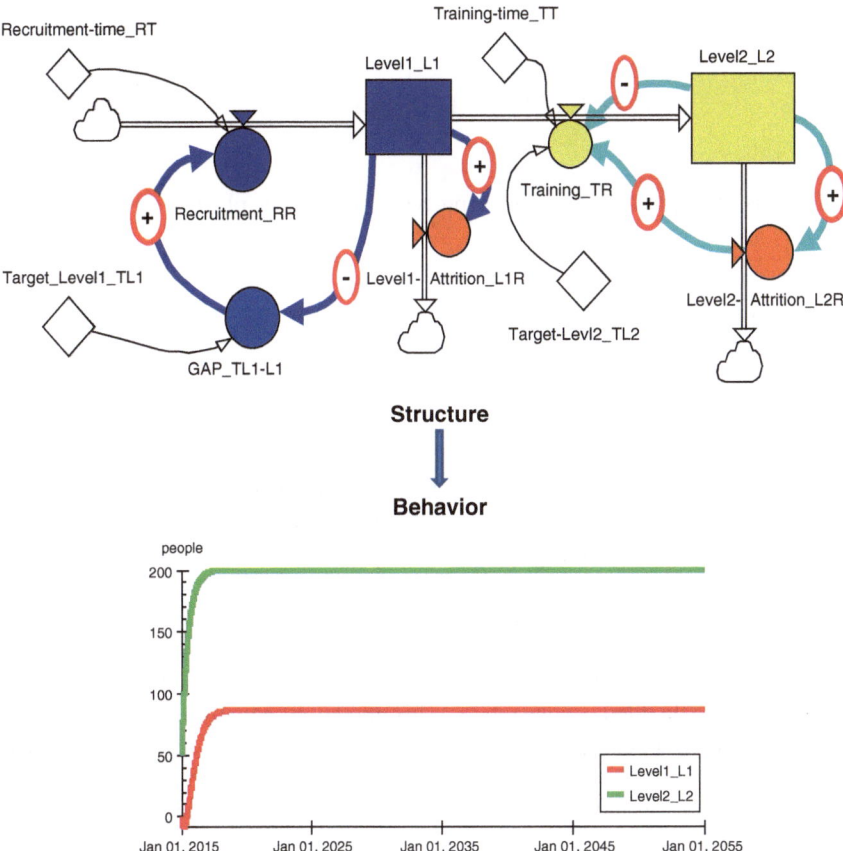

Fig. 5.6   Structure–behavior diagram of power plant's manpower

The recruitment rate (RR), depending upon the need (GAP = TL1 (target for level 1) − L1) and the time it takes for the recruitment process (RT), flows into L1 and the attrition rate (L1R) and the training rate (TR) are the outflows of L1.

The stock of trained people (e.g., nuclear power plant operators), L2, is fed by the training rate. The training rate (TR) depends on (i) available trainees, (ii) the stock L1, (iii) the training time TT, and (iv) the target level of trained personnel (i.e., how many trained personnel do we need?). The only outflow from L2 is the attrition rate, L2R. This level of disaggregation of manpower process is for the purpose of simplicity. Otherwise, one can add as many levels of disaggregation as the purpose of the model suggests. Also, note that here we have modeled a "pull" mechanism: personnel are hired and trained as per the requirements of the project. However, if the situation warrants a "push" mechanism, a similar structure can be developed. The mathematical model of these accumulations processes is given as:

$$L1(t) = \int_{t_0}^{t} [RR(s) - L1T(s) - L1R(s)]ds + L1(t_0)$$

$$L2(t) = \int_{t_0}^{t} [L1T(s) - L1R(s)]ds + L2(t_0)$$

For this example of a manpower process for a power plant, we have assumed that:

TL1 = 100 people.
RR = 20 people/year.
L1R = 0.1/year.
TL2 = 200.
TR = 15 people/year.
L2R = 0.1 /year.

By simulating this model, we can see the structure-behavior diagram[1] in Fig. 5.6. The level of trained people, L2, reaches the desired level of 200 people relatively faster than the level of newly recruited people reaches its desired level of 100 people. This difference is due to the fact that the net flow of L2 is higher than the net flow of L1, confirming yet again that it is the net flow that determines the state of any stock, a key principle of the physics of stocks and flows. Using this simple manpower planning structure, decision makers can monitor and control the level of their workforce in their organization. This model is equally applicable for the effective management of personnel across all the divisions of any energy system (e.g., mining, transportation, generation, transmission, and distribution). Likewise, if the purpose of the modeling activity is to delve into the process of manpower

---

[1] A structure-behavior diagram is a tool that links a stock and flow diagram of any process with its (simulated) behavior or output or performance.

attrition (i.e., understanding the effects of policy incentives/rewards, rules, norms, and culture), the system dynamics modeling approach is flexible enough to model such details.

5.4   Dynamics of Energy Production Process: Energy Supply System

To meet the energy demand of an economy, governments make substantial investments in both the production capital (e.g., power plants, machines, equipment, and technologies) and fuel resources (e.g., fossil and alternate fuels). In fact, energy supply mix is a critical issue of modern times. For example, what the optimal (based on varying criteria) electricity supply mix is for any nation or region is an ongoing policy issue across the globe. In the case of fossil fuel-fired power plants (i.e., oil, gas, and coal-fired power plants), still the dominant source of electricity supply in the world, onsite availability of fuel resources is critical. In the case of imports, the availability of fuel becomes even more uncertain and further complicates the design of sustainable energy policies.

Given that we have enough fuel resources available at the electricity production facility, the physics of stocks and flows of an electricity production process can be seen in Fig. 5.7. In this model, the electricity production depends on two factors: onsite availability of fuel resource, and available production capacity, subject to the constraints of capacity utilization. The capacity utilization of each of the power plants is a function of the ratio between the demand for electricity and the capacity to produce electricity. If this demand/supply ratio exceeds unity, then the capacity utilization function allows the power plant to run at 100 % capacity. But when this ratio drops below unity, the capacity utilization is reduced in proportion to demand. Production of electricity depletes the resource stock, so does the depreciation rate to the stock of production capacity. Mathematically, the electricity production process can be represented as

$$\text{Electricity Production Rate} = \text{Electricity Production Capacity}$$
$$* \text{Capacity Utilization},$$
$$\text{Electricity Production Capacity} = \text{MIN}(\text{Fuel Potential}, \text{Capital Potential})$$

where Fuel Potential = f(Resource, Resource Efficiency), and Capital Potential = f(Production Capital, Capital Productivity).

The structure–behavior diagram (i.e., Figure 5.6) is an actual result extracted from our recent study that analyzed the evolution of the electricity supply mix of Ontario, Canada (Qudrat-Ullah 2014). Although this result shows electricity

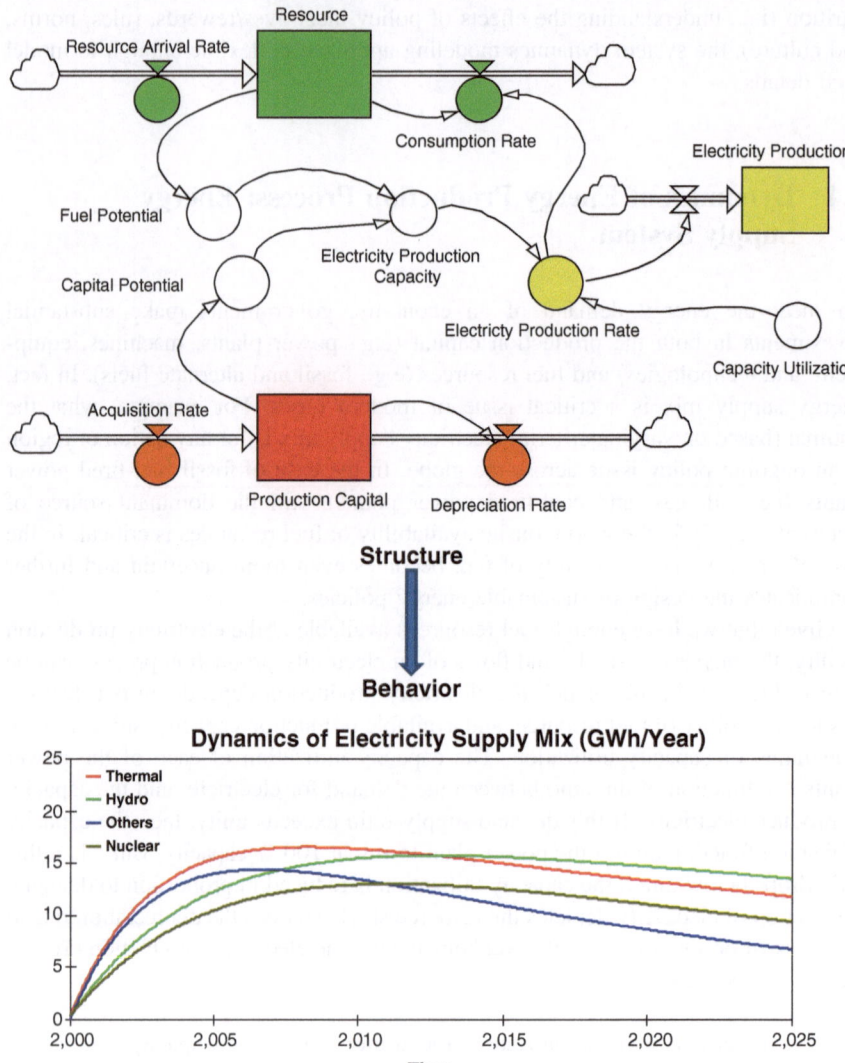

Fig. 5.7 Structure–behavior diagram of electricity supply mix

production from only four competing technologies, the structure of stocks and flows is quite generic and can accommodate any reasonable number of generation technologies.[2]

---

[2]The only limitation is the capability of the modeling packages/platforms in terms of their available dimensions one can use for any variable.

5.5 Stocks and Flows of Environmental Emissions and CO_2 Tax

The threats of environmental emissions (e.g., CO_2, SO_2, NO_x) and consequently global warming enter directly or indirectly in all energy policy decisions, be it in a public or private domain. The complexity of the issue has long attracted the attention of the system dynamics community. In fact, several researchers have implemented stock and flow based models (i.e., system dynamics models) to assess environmental effects of CO_2 emissions in energy systems (Feng et al. 2013; Trappey et al. 2012; Anand et al. 2005; Jin et al. 2009; Han and Hayashi 2008; Qudrat-Ullah 2013, 2014, 2015). If we want to control emissions, we need to understand the dynamics of various stocks and flows involved in the environmental emissions process. Again, to illustrate the physics of the relevant stocks and flows, we present a simplified, submodel of an actual model (Qudrat-Ullah 2005).

Figure 5.8 represents the substructure of the model that determines (endogenously) CO_2 tax and the realized environment premium for each of the generating technologies including the relevant feedback loops with their polarities (i.e., both the positive and negative feedback loops). Figure 5.9 presents the key structural elements, stocks and flows, of this system. The CO_2 tax is the amount of money spent (on account of abatement of CO_2 emissions) per unit of electricity produced. Except for hydro-based generation the electricity production causes CO_2 emissions. The target level of accumulated CO_2 emissions sets the amount of emissions to be treated (mitigated). The clean-up cost (CO_2 mitigation cost) determines the desired amount of investments for mitigating the set amount of emissions. The desired investments, together with the safety margin (for the mitigation program this is the time to spend taxes) determines the indicated level of tax income. The time to adjust the desired tax income, a little margin based on past spending, and the indicated tax income constitute the desired tax income.

The accumulated tax income specifies the spending limit (for the clean-up operating costs). The higher desired investments, the higher are the clean-up operating costs. The more spending (the clean-up operating costs) on treatment of the emission, the more is the CO_2 clean-up rate. Some of the CO_2 emissions escape beyond the national boundary over a period of time. Only the emissions accumulated within the national boundary are the target of the mitigation program. The clean-up costs will be incurred only by the remaining stock of electricity (this stock is represented for accounting purposes only) that causes the accumulation of CO_2 within the national boundary. Therefore, this amount of electricity, untreated for CO_2 emissions, is used to calculate the unit clean-up costs.

As the objective of this model was to evaluate the impact of the implementation of the CO_2 tax, Fig. 5.10 presents the comparative evolution of CO_2 emissions under both regimes: without CO_2 tax and with CO_2 tax. In this example, we can see that in the environment-oriented scenario total emissions of CO_2 were restricted to no more than 20.2 million tons/year after 2000.

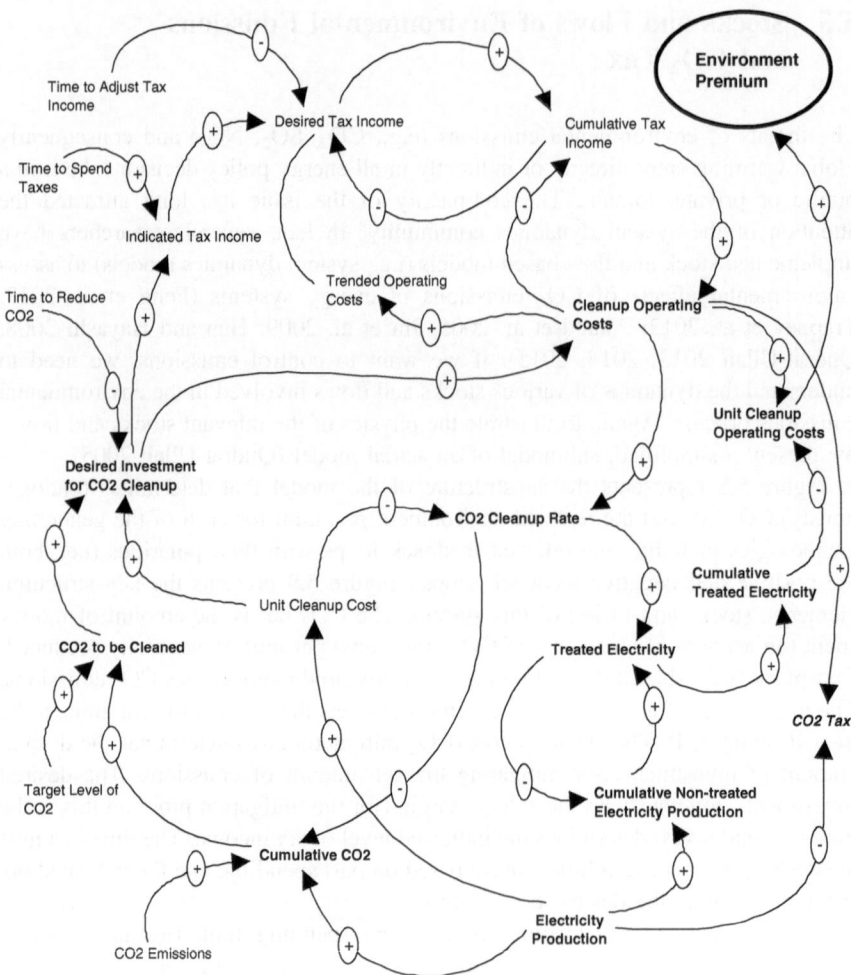

Fig. 5.8 A CLD of CO_2 emissions and CO_2 tax

Prior to the implementation of the environment-oriented policy mix, thermal capacity is steadily substituted being the least-expensive technology. Consequently, CO_2 emissions experience a linearly increasing accumulation in the environment. With a relatively small magnitude of 20.26 million tons (CO_2) in 1990, CO_2 emissions attain the largest increase of 250 % (compared to the 1990 levels) rather quickly in 2000 (Table 5.2). At this time, CO_2 emissions control policy is invoked. The imposition of control reduces the CO_2 emissions through the process of substitution for the least (or none) emitting electricity-generating technology as well as by improving (5 % in this scenario) the efficiency of power plants. Also, the CO_2 tax income (generated based on CO_2 emission levels and the corresponding mitigation costs) is spent over a period of time to treat the accumulated CO_2 emissions.

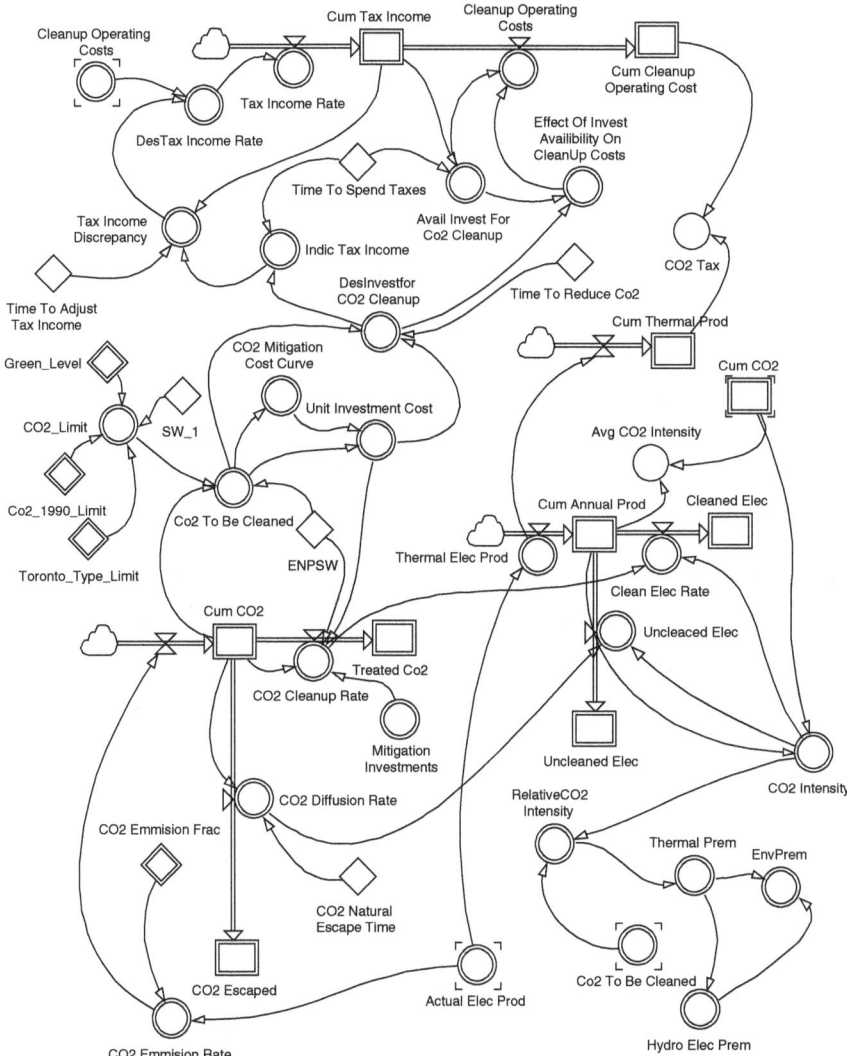

Fig. 5.9 A stock and flow diagram of CO_2 emissions and CO_2 tax

These control measures result in the reduced levels of CO_2 emissions. Eventually, we see that CO_2 emissions are under the binding constraint (a decrease of 2 % compared to the limiting value of 20.2 million tons/year) in 2009. After 2009, CO_2 emissions remain under the environmental constraint except for the last couple of years of the remaining time horizon. The delays involved in both tax spending and CO_2 emissions mitigation-capacity building cause a little increase in the level of CO_2 emissions for the later years of the time horizon (please see Table 5.2).

The thermal electricity generation produces CO_2 emissions. Based on the amount of emitted CO_2, the mitigation costs are realized for each of the technologies.

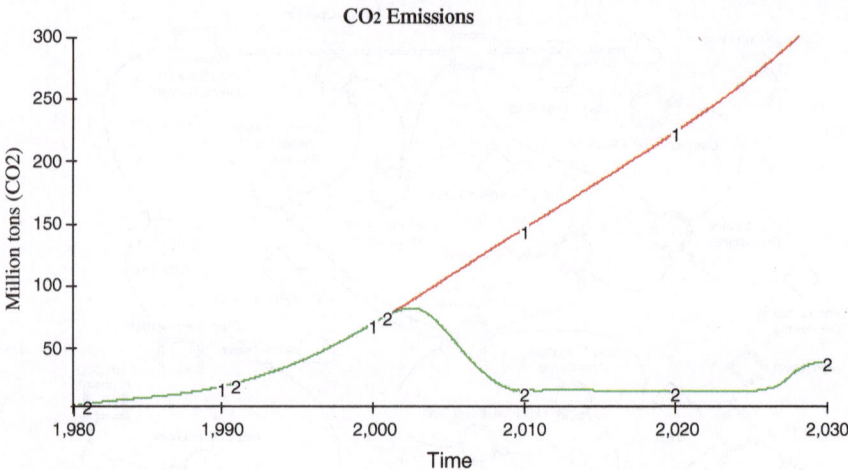

Fig. 5.10 Comparative (with and without CO_2 tax) evolution of CO_2 emissions

Table 5.2 Comparison of CO_2 emissions (with vs. without CO_2 tax regime)

Year	CO_2 emissions (million tons/year)		% change (compared to base case level)
	Base scenario	Environment scenario	
2000	70.98	70.98	0
2005	108.54	62.04	−43
2010	147.18	17.01	−88
2015	185.19	16.69	−91
2020	224.72	16.08	−93
2025	269.36	16.41	−94
2030	322.28	40.56	−87

Source Qudrat-Ullah (2005)

These mitigation costs form the basis for the CO_2 tax in our model and are shown in Fig. 5.11. Beginning in 2001 at 0.05 ($/MWh), CO_2 tax attains the maximum value of 0.22 in 2008. However, under the influence of the environment-oriented policy mix, the least-emitting technologies are preferred, leading to a substitution. But this substitution process remains active as long as the substitution is economically feasible. Consequently, the CO_2 tax stabilizes at a value of 0.22 $/MWh for the rest of the time horizon.

Although in this example we have elaborated on the physics of key stocks and flows related only to one type of emission, that is, CO_2 emissions, the structure presented here (i.e., Figs. 5.7 and 5.8) is generic enough to include other emissions (e.g., SO_2, NO_x, etc.) as well. By observing the evolution of CO_2 emissions under both the regimes (i.e., with CO_2 tax and without CO_2 tax), energy policy analysts, decision makers, and planners can experiment with various tax schemes to find the one with which they are comfortable. Thus, utilizing stocks and flows based

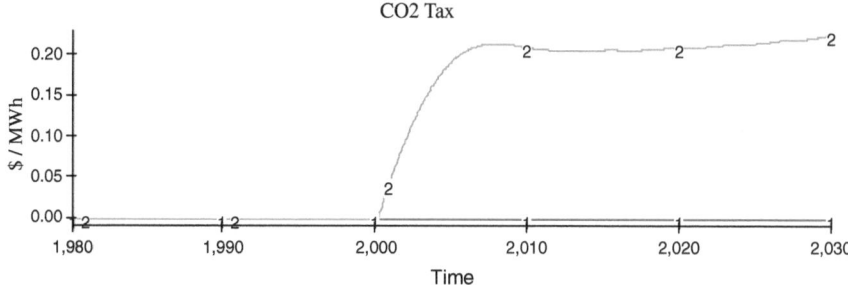

Fig. 5.11 Evolution of CO_2 tax under the environment-oriented scenario

modeling and analysis, one can benefit from the transparent nature of the methodology, allowing the users to "see" the structure that creates the dynamic "behavior" of the modeled process of the energy systems.

5.6   Understanding the Physics of Other Energy Systems Related Processes

Energy policy issues are diverse at best. We have elaborated on the physics of stocks and flows and the associated feedback structures of six critical and the most common processes of energy policy domain: energy demand, policy incentives and interfuel substitution mechanisms, production capacity, manpower recruitment and training, energy production, and environmental emissions and CO_2 tax, in Sects. 5.1–5.5 above. For the understanding of the physics of other energy systems related processes (e.g., resource, transportation, financial, regulatory, etc.), Table 3.1 provides a comprehensive list of studies that contain actual system dynamics models that embody such related structural processes. The interested reader can explore the developed system dynamics models of the respective thematic area.

5.7   Stocks and Flows Based View Is All About the Integration

By now, when we have introduced and discussed various energy system processes, it has become clear that: (i) energy system processes are a web of various feedback loop structures (e.g., see Figs. 5.10 and 5.11), and (ii) these interacting feedback loops generate the outcome or behavior of the process.

Therefore, when our objective, say is "to design a sustainable energy policy for a nation, region, or a country," we are essentially looking into a systematic and integrative process to design a system of interacting parts (Fig. 5.12) whereby:

(i) the objectives (issues) are well defined;

(ii) causal relationship with in and across various levels of actors and resources of the system are identified and are represented by casual loop diagrams;

(iii) the physics of the system (i.e., the movement of material and the influence of information) is represented by stocks and flows with explicit modeling of associated uncertainties, nonlinearities, and time lags;

(iv) utilizing case-specific data and/or relevant theoretical and empirical support, simula ted behavior (of modeled structure) is achieved;

(v) what-if scenario discovery analysis is entertained, insights gained, and decisions are made (and implemented).

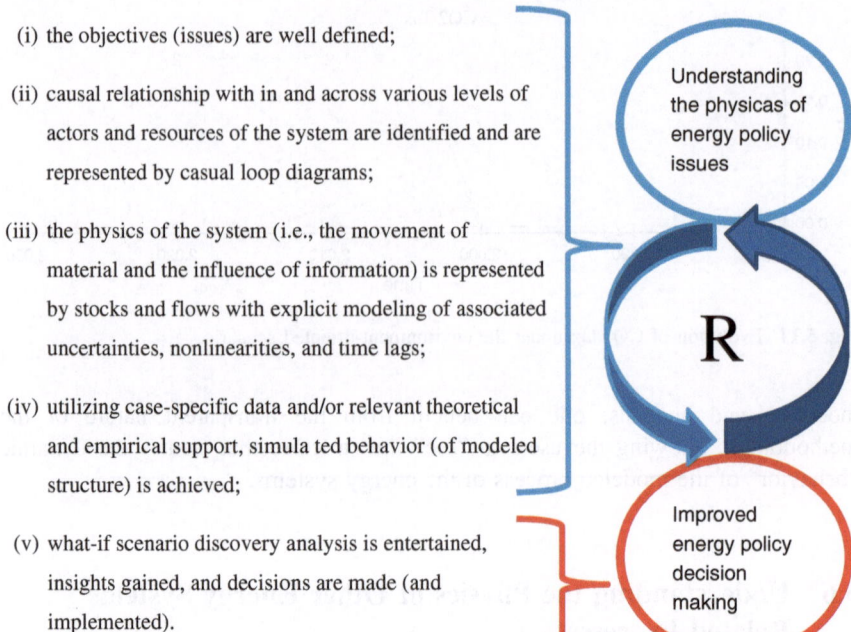

Fig. 5.12 The virtuous feedback loop (any feedback loop that is favorable to the goal or objective of the systems is called a virtuous feedback loop) between understanding and decisions

Figure 5.12 provides an overview of the stock and flow perspective; it encourages a learning-action orientation towards the understanding of complex, dynamic energy policy issues: it allows the decision makers to embrace an improved decision-making approach by understanding the physics of underlying feedback loop structures. Thus, the key thrust of the stock and flow perspective to address complex dynamic issues of the energy policy domain is that:

(i)   The sum of parts is not equal to the whole (e.g., having two power plants with the capacity of 300 MW each does not mean that 600 MW capacity will be operational and all the time!).

(ii) "Seeing the feedback" is necessary but making a habit of using it in decisions will be of real help.

(iii) It is the interacting feedback loops that run the show (i.e., they generate the behavior, good or bad); it is an absolute necessity (for the decision makers in energy policy domain) to see the big (holistic) picture.

5.8 Summary

In this chapter, we set the objective of understanding the physics of stocks and flows (i.e., the interaction of feedback loop structures) of some fundamental processes of energy systems including energy demand, policy incentives and interfuel substitution mechanisms, manpower recruitment and training, energy production, and environmental emissions and CO_2 tax. We attempted to achieve our objective by making use of various tools and techniques of system dynamic modeling: causal loop diagrams, stock and flow diagrams, mathematical formulations, simulated outputs of some actual submodels of existing system dynamics models, and structure–behavior diagrams. In particular, we illustrated the core principle of system dynamics; structure drives its behavior, with the help of several actual examples (where data come from real, published case studies). With such an understanding of feedback loop structures and structure–behavior relationships about the various energy systems processes, the energy policy decision makers are expected to develop insights and make better decisions. In the next chapter, we showcase some examples of successful development and applications of stock and flow based (i.e., system dynamics) models: the physics of stocks and flows of energy systems in action.

References

Abada, I., Briat, V., & Massol, O. (2013). Construction of a fuel demand function portraying interfuel substitution, a system dynamics approach. *Energy, 49*(1), 240–251.

Anand, S., Vrat, P., & Dahiya, R. (2005). Application of a system dynamics approach for Assessment and mitigation of CO_2 emissions from the cement industry. *Journal of Environmental Management, 79*, 383–398.

Feng., Y., Chen., Q., & Zhang, X. (2013). System dynamics modeling for urban energy consumption and CO_2 emissions: A case study of Beijing-China. *Ecological Modelling, 252*, 44–52.

Ford, A. (1996). System dynamics and the Electric Power Industry. *System Dynamics Review, 13*(1), 57–85.

Han, J., & Hayashi, Y. (2008). A system dynamics model of CO_2 mitigation in China's inter-city passenger transport. *Transportation Research Part D: Transport and Environment, 13*(5), 298–305.

Jin, W., Xu, L., & Yang, Z. (2009). Modeling a policy making framework for urban sustainability: Incorporating system dynamics into the ecological footprint. *Ecological Economics, 68*(12), 2938–2949.

Moxnes, E. (1990). Interfuel substitution in OECD–European electricity production. *System Dynamics Review 1990, 6*(1), 44–65.

Ontario Energy Board (OEB) (2002). Ontario's System-Wide Electricity Supply Mix 2002. http://www.ontarioenergyboard.ca/documents/electricity_mix.pdf. Accessed on July 22, 2015.

Qudrat-Ullah, H. (2005). MDESRAP: a model for understanding the dynamics of electricity supply, resources and pollution. *International Journal of Global Energy Issues, 23*(1), 1–13.

Qudrat-Ullah, H. (2013). Understanding the dynamics of electricity generation capacity in Canada: A system dynamics approach. *Energy, 59*, 285–294.

Qudrat-Ullah, H. (2014). Green power in Ontario: A dynamic model-based analysis. *Energy, 77* (1), 859–870.

Qudrat-Ullah, H. (2015). Independent power (or pollution) producers? Electricity reforms and IPPs in Pakistan. *Energy, 83*(1), 240–251.

Saeed, K. (2013). Managing the energy basket in the face of limits: A search for operational means to sustain energy supply and contain its environmental impact. In H. Qudrat-Ullah (Ed.), *Energy policy modeling in the 21st Century* (pp. 69–86). New York: Springer.

Trappey, A., Trappey, Ch V, Gilbert, L., & Chang, Y. (2012). The analysis of renewable energy policies for the Taiwan Penghu island administrative region. *Renewable and Sustainable Energy Reviews, 16*(1), 958–965.

Chapter 6
Physics of Stocks and Flows in Action

> Experience is the past tense of experiment
> —Gregory Alan Elliot

As we have presented earlier in Chap. 2 of this brief book (please see Table 2.1), a fairly large number of energy policy models have been developed by using the system dynamics approach over a period of about five decades (Qudrat-Ullah 2015). Here, in this chapter, our objective is to present a snapshot view of some selected models[1] to see how these stocks and flows based models have contributed to the better understanding of various energy policy related issues. Specifically, we describe: (i) the context and objective (i.e., who and why a model was developed), (ii) a brief overview of model structure (i.e., identification of its key structural elements), (iii) the validity of the model (i.e., how the validation, if any, of the model was conducted), and (iv) key insights and major policy implications.

6.1 Model 1: Green Power in Ontario (Source: Qudrat-Ullah 2014)

6.1.1 Context and Objective of Model (i.e., Model 1) Development

Here is the context and the objective of development of Model 1:

Since 2005, the situation of electricity supply in Ontario, Canada's largest province by both population and electricity generation, has changed. The Ontario Power Authority (OPA) was given the mandate to ensure reliable, sustainable and cost-effective supply system to avoid any shortage of electricity in Ontario. In 2007, OPA prepared a long-term plan, Integrated Power System Plan (IPSP). In 2009, government enacted Green Energy Act that resulted in numerous "demand side management developments", "conservation initiatives", and "renewable energy projects". Consequently, in 2013, Ontario has "surplus" generation

---

[1] All these models are published in peer review outlets and are publicly available.

© The Author(s) 2016
H. Qudrat-Ullah, *The Physics of Stocks and Flows of Energy Systems*,
SpringerBriefs in Complexity, DOI 10.1007/978-3-319-24829-5_6

capacity (OPA 2013; CleanAA 2012). Now with excess generation capacity regime, Ontario's electricity sector presents several challenges as well emerging opportunities.

The key dilemma in Ontario's electricity system is that although it has excess supply, its effect on electricity rates are in opposite direction—consumers are continuing to pay higher electricity rates. At the outset, one could argue that for relatively more green power, you have to pay higher prices. But in reality, there is limit to even much needed "belt tightening"—consumers are willing to pay only so much and for so long. In fact, a recent study has shown an increasing decline in the support of regulated renewable energy initiatives in Ontario (Stokes 2013). Others argue that issue is with Ontario's electricity supply mix: with a more balanced supply mix with system-wide considerations of socio-economic and environmental effects, similar green power share could be sustained at better competitive prices (Stokes 2013; Ross 2013).

The problem of aging infrastructure of electricity sector coupled with uncertain fuel prices, and unpredictable cost and performance of relatively new demand side initiatives adds complexity of Ontario's electricity system. Therefore, the challenge is not only to avoid the possibility of any potential imbalances in electricity sector of Ontario, but also create and maintain a balanced and cost-effective supply mix that fully accounts for the dynamics associated with socio-technical, economic, environmental factors.

On the other hand, with no fear of potential electricity-shortage related blackouts at hand, Ontario's electricity sector planers and decision makers can avail the opportunity to assess the Ontario's current electricity supply mix and design and craft an alternate balanced and sustainable (i.e., within the constraints of socio-technical, economic, and environmental factors) supply mix. The objective of this study, therefore, is to contribute in this area by (i) offering a review of the dynamics of electricity sector, and (ii) providing an integrated, system-wide, simultaneously accounting for both the supply and demand side's key factors, analyze the dynamics of Ontario's electricity system. We will achieve this objective by investigating: **Within the constraints of socio-technical, economic and environmental concerns what would be the best combination of electricity supply mix and demand side initiatives to achieve sustainable electricity (Green Power) in Ontario?** (Qudrat-Ullah 2014).

6.1.2 Overview of Model 1 Structure

Figure 6.1 presents an overview of key sectorial interactions (i.e., the key feedback loop structures) of the dynamic model. We can trace around this diagram to find that this model focuses on Ontario's supply-mix and demand-side initiatives and their interactions with pricing, profitability, investments, and electricity production sectors of the electricity system. Both positive and negative feedback loops are present with the explicit representation of a construction delay (T_i) for each technology. The interaction of these feedback loops will create the behavior to be analyzed for aiding decision makers to gain insights and make better decisions.

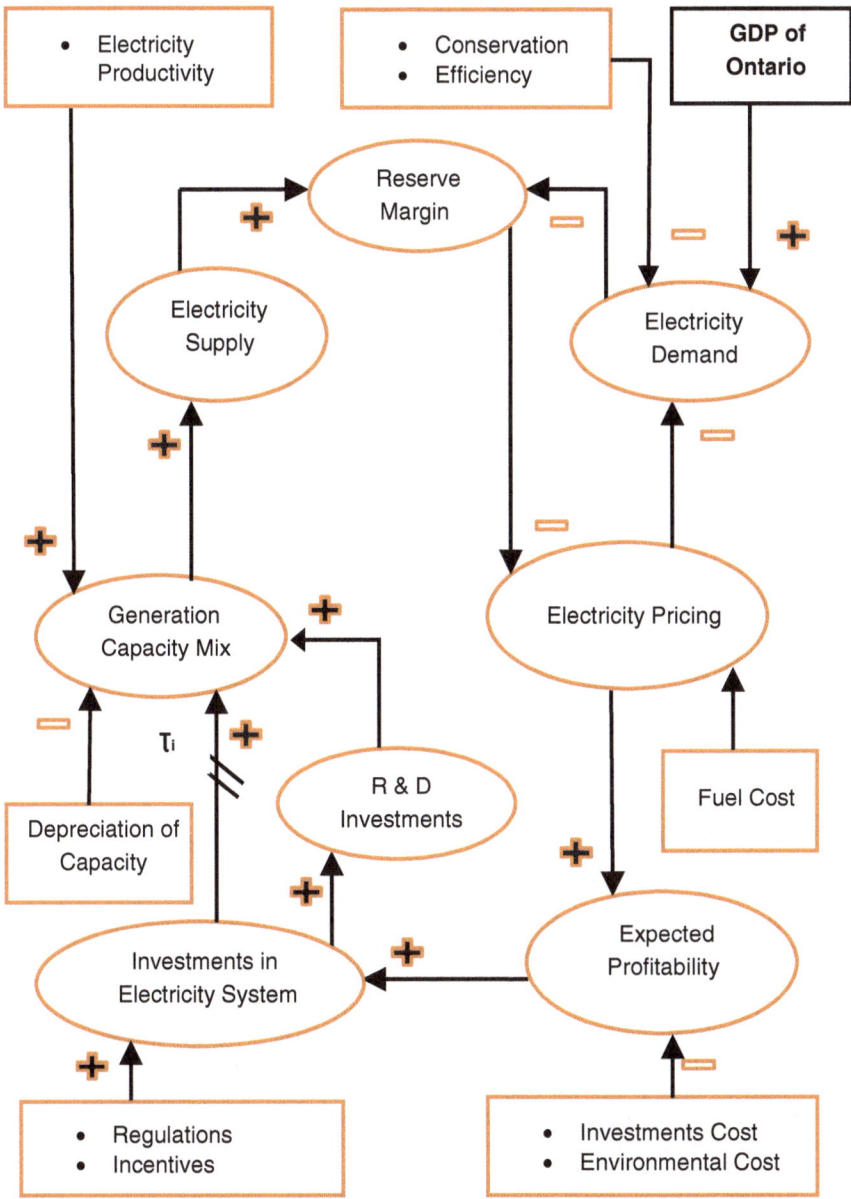

Fig. 6.1 Overview of sectorial interactions in the dynamic model, Model 1. *Source* Qudrat-Ullah (2014)

6.1.3   Validity Testing of Model 1

Before any model is put to its use, it is imperative that it is valid (i.e., it has passed the relevant validity tests). System dynamics models are causal models (Forrester 1961; Barlas 1989; Sterman 2000). They are well suited for policy analysis and assessment rather than the point prediction of the variables under study. In fact, in system dynamics, the validation is the process of building confidence in the usefulness of a model (Forrester 1961; Sterman 2000; Qudrat-Ullah and BaekSeo 2010). Both the structural and behavior validity procedures are applied to system dynamics models. Although structural validity ascertains that model structures generate the right behavior, behavior validity assesses how well the model-generated behavior mimics the observed patterns of the real system (Barlas 1989). Our model, Model 1, was successfully evaluated against all these procedures (for details, please see Qudrat-Ullah and BaekSeo 2010 and Qudrat-Ullah 2014).

6.1.4   Key Insights and Energy Policy Implications

The validated model, Model 1, was applied to evaluate three scenarios (Qudrat-Ullah 2014):

(i) *Status Quo Scenario (SQS)* serves as the reference scenario. It is based on Ontario's economic data and conditions.
(ii) *Renewables-Focused Scenario (RFS)* simulates what would happen if there is a larger support for renewables in the electricity supply mix of Ontario.
(iii) *Low-Carbon Economy Scenario (LES)* focuses on electricity generation from the technologies with the least CO_2 emissions.

Based on the results of Model 1, several important implications regarding the dynamics of Ontario's electricity system are drawn:

- With "excess capacity" regime at hand, form now to the end of this decade, 2020, Ontario has the opportunity to reconsider OPA's plan. For instance, our results show that "Dash for gas" approach will neither help in lowering the electricity rates nor it will reduce the stock of cumulative electricity related CO_2 emissions. Instead, substitution of relative high cost nuclear refurbishments with low cost hydro and nuclear solutions could result in similar CO_2 emissions accumulations and intensity but at relatively cheaper "green power."
- If Ontarians are willing to pay higher electricity rates (i.e., increase of 11 % per year from 2020 to 2030) for (i) having more renewable generation (i.e., 30 % share in the supply mix) and (ii) a reduction of 14 % CO_2 emission intensity both in 2020 and 2030 then staying the course with OPA's plan is not a viable option. A new plan that is able to achieve these results, like our RFS, should be actively pursued by the decision makers.
- If Ontarians prefer their future to have more low-carbon economy than what is expected under OPA's plan, then there is good news: affordable plan can be crafted. With a focus on the reduction of thermal generation, addition of renewable generation, and

investments in R&D of electricity system (i.e., improving on electricity productivity and power plants' operational efficiency gains), more low-carbon economic and industrial activity can be achieved in an affordable way (i.e., with the availability of relatively cheaper electricity). Similar assertions are advanced by others as well (Pollution Probe 2012). In addition, such a plan can also make Ontario's supply mix import- independent.

- Despite being a significant source of CO_2 emissions, OPA's plan favors gas-based generation for its status as a "flexible supply" utility. Based on our analysis, however, utilization of hydro reservoir as "energy storage systems" appears to be a cheaper source of "flexible supply" than the gas-based generation. In fact, if such a role of hydro is implemented, gas-generation can be eliminated from the supply mix. Thus, resulting in more and affordable "green power" for Ontarians.
- Currently, Ontarians can have more renewable generation, other than hydro, but with an increasing electricity rates for a long time. On the other hand, based on increased productivity and enhanced efficiency resulting from R&D investments, renewable generation can have similar share in the supply but with relatively cheaper rates of electricity during the same time period.
- In contrast to current OPA's plan where electricity consumption per capita in Ontario still remain higher than comparable states like New York, a new plan focused on having more low-carbon economy seems to lower the electricity intensity. Lower electricity intensity in Ontario means more "green power." (Qudrat-Ullah 2014)

6.2   Model 2: MDESRAP (Source: Qudrat-Ullah 2015)

6.2.1   The Context and Objective of Model (i.e., Model 2) Development

The context and the objective of development of the model, MDESRAP as reported by the author (myself) is:

Like other countries in the world, Pakistan started liberalization and privatization of its energy sector in early 1990s. In response to Pakistan's 1994 Energy Policy that provided attractive incentives to private investors, offers of over 4000 MW electricity generation capacity by independent power producers (IPPs) were realized. Key incentives of this policy were [18]:

- Low taxes, duties, and fees
- Power purchase guarantee (bulk power tariffs) @ 6.5 cents (US)/KWh
- Fuel supply arrangements
- Foreign exchange risk insurance

Despite these lucrative incentives, electricity supply sector of Pakistan has been struggling to meet the electricity demand. In fact, beginning in 2005-2006 Pakistan is in a crisis with regards to the supply of electricity. Pakistanis were enduring as much as eighteen hours per day of load shedding (i.e., no electricity at all) for months [22]. The gap between demand and supply of electricity peaked at 6000 MW in 2012 [6, 21]. This worst situation of electricity shortage has not improved even today in the summer of 2014 [45]. Besides the obvious failure of the 1994 Energy Policy of Pakistan in meeting the demand of electricity, none of the IPPs invested in the country's rich endogenous resource— hydroelectricity. Instead, they invested mostly in gas and oil fired power plants. In fact, IPPs are known to invest in relatively early-return producing technologies, a "dash for gas" phenomenon [20, 31, 36, 46]. Thermal generation, on the other hand, has severe implications for the environment of Pakistan.

With the ongoing sufferings of people and the economy of Pakistan due to severe shortage of electricity, Pakistan's energy sector planers and decision makers need a comprehensive assessment and analysis of existing energy policy which is a modified version of 1994 Energy Policy. In particular, with the available empirical evidence of mixed performance of IPPs, both in the developed and developing nations, [20, 46], there is a compelling need for systematic evaluation of IPPs' performance in the electricity sector of Pakistan across socio-technical, economic, and environmental dimensions. Specifically, the long-term impact of IPPs investment on the host country's environment (e.g., environmental pollution due to electricity generation related CO_2 emissions) is rarely assessed. Likewise good governance and ethical practices that IPPs have shown in the case of developed nations are less often executed whey they invest in developing countries [31].

The objective of this study, therefore, is to contribute in this area by providing an integrated, system-wide simultaneously accounting for both the supply and demand side's key factors, analysis of the dynamics of Pakistan's electricity system. **We will attempt to achieve this goal by addressing the research questions: How does IPPs' investment in Pakistan's electricity sector impact the emissions of CO_2 in the environment? Within the constraints of socio-technical, economic and environmental concerns, what would be the best combination of electricity supply mix and supply side initiatives to achieve sustainable electricity supply in Pakistan?** (Qudrat-Ullah 2015)

6.2.2 Overview of MDESRAP Structure

MDESRAP[2] is organized into seven sectors, namely electricity demand, investment, capital, resource, production, environment, and costs and pricing sectors, as shown in Fig. 6.2. The interaction of the feedback loops of one sector to the other leads automatically to the closure of feedback loops that govern the behavior of the system (Qudrat-Ullah 2005).

6.2.3 Validity Testing of MDESRAP

As with any other system dynamics model, MDESRAP was subjected to various structural and behavioral tests including Theil's inequality statistics and it passed all the tests satisfactorily (please see details in Qudrat-Ullah 2004). On the utility of MDERAP, an anonymous reviewer commented in 2014 that:

> ... In 2001, Professor Qudrat-Ullah was successfully able to forecast the adverse impact of policy [i.e., the 1995 energy policy of Pakistan], which shifted the energy mix by moving electricity generation to more polluting sources, i.e., oil and gas. He forecasted higher

---

[2]MDESRAP was originally developed in 2001 (please see the details in Qudrat-Ullah and Davidsen 2001).

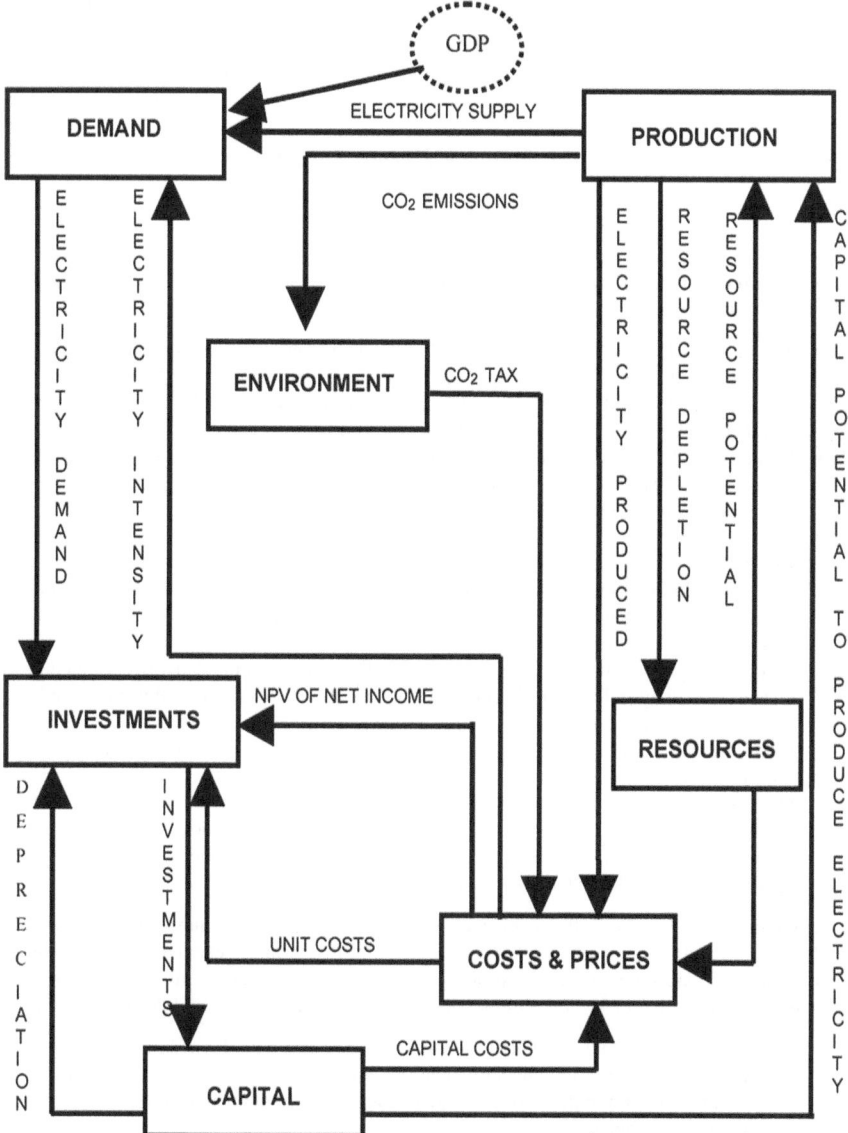

Fig. 6.2 Feedback loop interactions in MDESRAP. *Source* Qudrat-Ullah (2005)

emission level and increased reliance on oil, which had to be imported, thus increasing the import bill.

The energy crises in Pakistan peaked in 2010 and it unfolded exactly as Professor Qudrat-Ullah had forewarned in 2001. I believe that had Professor Qudrat-Ullah's paper been read by policymakers in Pakistan and had they choses to respond to the concerns he has raised, Pakistan would have been able to deal with energy as well as environmental crisis in a more effective way.

Such accolades from external peers certainly add to the methodologically tested validity of our dynamic model, MDESRAP.

6.2.4 Key Insights and Energy Policy Implications

Based on the results of MDESRAP,

several important insights and implications regarding the dynamics of Pakistan's electricity system were drawn:

- The existing Energy Policy fails on at least four aspects: (i) it is not environmental friendly, (ii) the electricity rates appears to rise for a long-period, (iii) the gap between the demand and supply of electricity does not appear to close in a balanced manner, and (iv) it increases dependence on imported furnace oil.
- Contrary to the common pitch for the benefits of privatization, IPPs produce neither cheaper nor cleaner electricity in Pakistan. Instead, they are the dominant contributors to the electricity-related CO_2 emissions in Pakistan.
- With the "electricity crisis" at hand, Pakistan has the opportunity to reconsider the Energy Policy. For instance, our results show that the IPPs' approach, the dominant investment in quick-money-generator pro-oil generation, will neither help in lowering the electricity rates nor will it reduce the stock of cumulative electricity related CO_2 emissions. Instead, substitution of indigenously resource-rich and relatively low-carbon hydro and nuclear solutions, such as in our AEPS, could result in relatively cheaper and cleaner electricity in Pakistan. In addition, such a plan can also make Pakistan's supply mix import independent.
- Despite being a significant source of CO_2 emissions, the Energy Policy favors pro-oil thermal generation for its status as a "flexible supply" utility. Based on our analysis, however, utilization of hydro reservoirs as "energy storage systems" appears to be a cheaper source of "flexible supply" than thermal generation. In fact, if such a role of hydro is implemented, thermal generation can be eliminated from the supply mix. Thus, resulting in more affordable "low-carbon electricity" for Pakistanis.

In additions to (a) a systematic review and evaluation of the Energy Policy of Pakistan, and (b) the presentation of an alternative energy plan with insightful, integrated analysis of the dynamics of Pakistan's electricity system in socio-economic and environmental dimensions, this study also makes two important contributions of the methodological nature:

(i) Our validated dynamic simulation model can serve as the underlying simulation model for the development of an interactive learning environment (ILE). The developed ILE can be a wonderful training tool for academicians and practitioners to assess and create viable energy policies.
(ii) We have presented an example of using "Thiel's Inequality Testing" to validate our dynamic model, our modeling and simulation community can make use of this approach to increase the validity and acceptance of their simulation models. (Qudrat-Ullahm 2015)

6.3   Model 3: Electricity Generation Capacity in Canada (Source: Qudrat-Ullah 2013)

6.3.1   The Context and Objective of Model (i.e., Model 3) Development

Here is the context and the objective of development of Model 3:

> The electricity production industry in Canada consists of four major nonrenewable sectors, and two major renewable sectors. The majority of nonrenewable electricity comes from crude oil, coal, natural gas, and uranium. On the other hand, the majority of electricity from renewable sources comes from hydroelectric and wind production. The relationship between supply and demand of electricity has changed over the past few economic cycles. It is important to note that conventionally, as the demand for electricity increased, the production of electricity also increased. A notable change came with the 1989-1993 recession, when the demand growth for electricity stalled. With the stall of demand, came the stall of supply (IFC 2006). However, as the economy recovered from the recession, demand growth resumed, however supply did not follow. Instead, the focus on maintaining alignment between supply and demand was on productivity. Demand is driven by increase in electricity using economic activities, and efficiency gains. Productivity may be further divided into mechanical efficiency and conservational efficiency or electricity spent for value addition. The two may be further divided into current machinery efficiency improvements, the invention of more efficient machinery, and the devising of new techniques that improve the value adding capabilities of processes. The driver behind such productivity improvements is research and development, which in turn is driven by investment (Park et al. 2007) and (Kilanc and Or 2008). For renewable energy sources, technological efficiency does not depend on the demand side dynamics directly. However, the economics and cost competitiveness of the technologies do (IFC 2006).
>
> Despite considerable improvements in the productivity area, Canada's electricity supply and demand system has experienced significant imbalance in recent history (IFC 2006). In fact, complexity of the system makes sustainable policy decision making a difficult task. Complexity of this system primarily comes from the existence and interactions of nonlinear and dynamic variables including various stocks of electricity generation capacity, restricting and regulatory regimes, fuel supply and price dynamics, and advances and challenges in technologies for electricity generation, transmission, and consumption. Therefore, **the key objective is to understand the causal relationships among the system's nearly immeasurable amount of variables and the patterns that exist in Canada's electricity system is essential for systematic and sustainable policy decisions**. (Qudrat-Ullah 2013)

6.3.2   Overview of Model 3 Structure

Model 3 is shown in Fig. 6.3. It focuses on investments in the electricity system and their interactions with capital assets and electricity production sectors via its feedback loops (Qudrat-Ullah 2013).

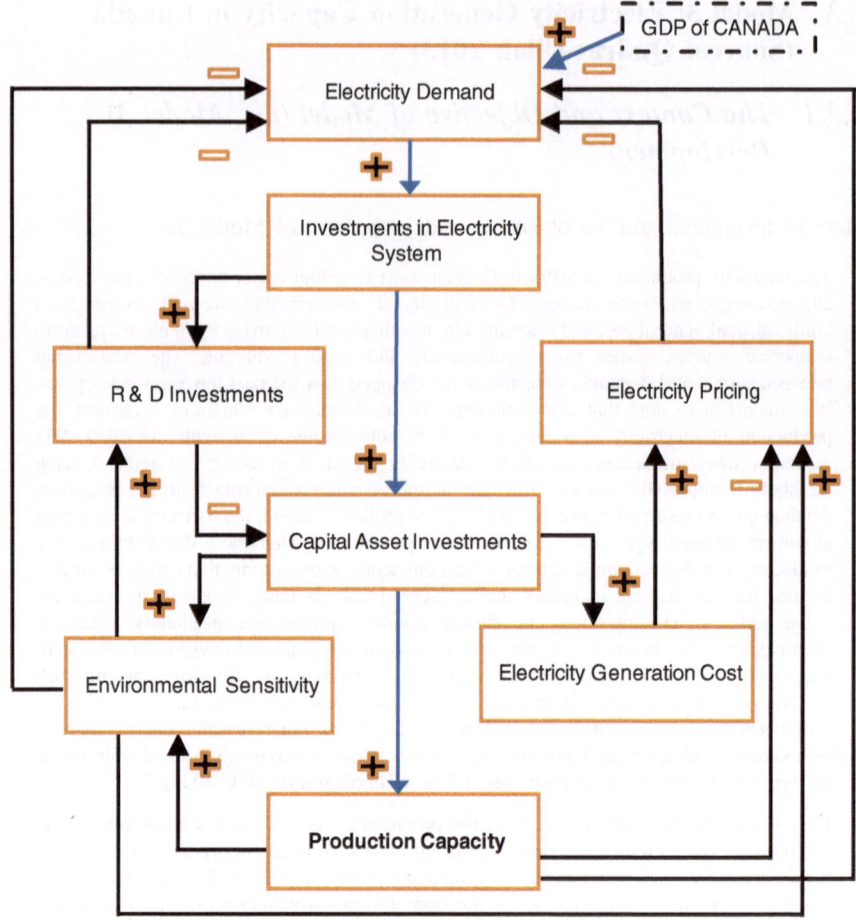

Fig. 6.3 Overview of feedback loop interactions in Model 3. *Source* Qudrat-Ullah (2013)

6.3.3   Validity Testing of Model 3

We rigorously applied both the structural and behavioral tests for the validation of our model. For structural validity, Model 3 was exposed to the following tests suggested by Barlas (1989) and demonstrated in Qudrat-Ullah et al. (2012):

(i)   *Boundary adequacy* test (i.e., whether the important concepts and structures for addressing the policy issues are endogenous to the model)

(ii) *Structure verification* (i.e., whether the model structure is consistent with relevant descriptive knowledge of the system being modeled)

(iii) *Structure verification* (i.e., whether the model structure is consistent with relevant descriptive knowledge of the system being modeled)

(iv) *Dimensional consistency* (i.e., whether each equation in the model dimensionally corresponds to the real system)

(v) *Parameter verification* (i.e., whether the parameters in the model are consistent with relevant knowledge of the system)

(vi) *Extreme condition* (i.e., whether the model exhibits a logical behavior when selected parameters are assigned extreme values)

For the behavioral validity of Model 3, we applied several behavioral tests (Barlas 1989; Sterman 2000) including:

- *Trend comparison and removal*: This test provides the relative best fit of data by estimating the trend of both the model-generated and the reference data series.
- *Autocorrelation function test for period comparison*: The autocorrelation test is used to detect the significant errors in the periods (e.g., time lags).
- *Cross-correlation function test for phase lag detection*: The cross-correlation function between the two time patterns shows how the two behavior patterns are correlated at different time lags.
- *Comparing the means*: This test examines the "Percent error in the means" to see the discrepancy between the means.
- *Comparing the amplitude variations*: This test compares the variations in the simulated output by computing the "Percent error in the Variations (*E2*)" as: $E2 = |SS - SA|/SA$, where SS and SA are the standard deviations of the simulated and reference time patterns respectively.

Model 3 passed all these tests satisfactorily and was used to perform several iterations of what-if discover scenarios.

6.3.4   Key Insights and Energy Policy Implications

Based on the results of Model 3:

several important insights and implications regarding the dynamics of Canada's electricity generation capacity were drawn.

With our theoretical review, dynamic hypothesis, and simulation model-based scenarios, we have attempted to explain the dynamics of variables acting within the electricity supply and demand system of Canada. Specifically, we have looked at variables within our generation capacity system. The key to the avoidance of a gap between electricity supply and demand, as well as sustainable, safe, and cost competitive production is to take advantage of the identified factors and potential policy decisions. In addressing our current supply and demand gap issue, we must not only continue to invest in capital assets for electricity production, but also continue our increased investments in R&D and productivity initiatives. Demand management and reduction, as well as production and end use machinery efficiency play prominent roles in maintaining stability throughout the system. Canada must be prepared to diverge from traditional adjustment methods and adopt new strategies focused on capital assets, productivity, and efficiency in order to avoid a downward spiral of electricity industry deficiency.

As per our model-based analysis, an additional investment of about 10 Billion $ over a decade (2015-2025) will not only allow Canada to effectively close the supply and demand gap but in a relatively greener way. With these additional investments, Canadian economy can also expect better energy intensity (.25 versus .21 (toe/Million$). This will result in wider recognition of Canada as a Green Economy (CBC 2010).

By utilizing our developed simulation model, future research can investigate other related issues in the context of alternative policy design for Canadian electricity sector. For instance, in the identified capacity-mix, which capital asset should be preferred the most? Our developed model is flexible enough to be adapted to model and analyze such issues. Therefore, besides providing useful policy insights on electricity generation capacity dynamics in Canada, this research contributes with an effective policy analysis and design tool in the form of a unique system dynamics based simulation model. (Qudrat-Ullah 2013)

6.4 Summary

As they say, "Seeing is believing," in this chapter we have provided a brief overview of three models: Model 1, Model 2 (MDESRAP), and Model 3; each was built to serve a particular purpose, yet all addressed the important energy policy issues. They served three different geographical areas: Ontario, Pakistan, and Canada.

Now, when you have gone through each of the models' context and objectives, feedback structures, validity procedures, and major insights they provided, it is not an unreasonable assumption that you have seen at least a glimpse of "the physics of stocks and flows in action," the major objective of this chapter. Therefore, we can assert that the stocks and flows based models, which are causal models, are well suited for energy policy analysis and design.

References

Barlas, Y. (1989). Multiple tests for validations of system dynamics type of simulation models. *European Journal of Operational Research, 42*(1), 59–87.

CBC. (2010). Conference Board of Canada. Energy intensity 2010. http://www.conferenceboard.ca/hcp/details/environment/energy-intensity.aspx. Accessed on May 2, 2013.

CleanAA. (2012). CleanAirAlliance.org. Ontario electricity surplus: An Opportunity to Reduce Costs 2012. http://www.cleanairalliance.org/files/surplus.pdf. Accessed on January 15, 2013.

Forrester, J. (1961). *Industrial Dynamics*. Cambridge, USA: MIT Press.

IFC Consulting. (2006). The electricity supply/demand gap and the role of efficiency and renewables in Ontario, 2006. http://www.pollutionprobe.org/old_files/Reports/elec_supplydemandICF.pdf. Accessed December 01, 2012.

Kilanc, P., & Or, I. (2008). A decision support tool for the analysis of pricing, investment and regulatory processes in a decentralized electricity market. *Energy Policy, 36*(8), 3036–3044.

OPA. (2013). Ontario Power Generation. Energy efficiency 2013a. http://www.opg.com/safety/efficiency.asp. Accessed on July 10, 2013.

Park, J., Ahn, N.-S., Yoon, Y.-B., Koh, K.-H., & Bunn, D. W. (2007). Investment incentives in the Korean Electricity Market. *Energy Policy, 35*(11), 5819–5828.

Pollution Probe. http://www.pollutionprobe.org/old_files/Reports/elec_supplydemandICF.pdf. Accessed on December 01, 2012.

Qudrat-Ullah, H. (2004). Resources, pollution, and development of sustainable energy policies. In M. A. Quaddus, & M. A. B. Siddique (Eds.), A handbook of sustainable development planning: Studies in modelling and decision support. UK: Edward Edgar.

Qudrat-Ullah, H. (2005). MDESRAP: a model for understanding the dynamics of electricity supply, resources and pollution. *International Journal of Global Energy Issues, 23*(1): 1–13.

Qudrat-Ullah, H. (2013). Understanding the dynamics of electricity generation capacity in Canada: A system dynamics approach. *Energy, 59*, 285–294.

Qudrat-Ullah, H. (2014). Green power in Ontario: A dynamic model-based analysis. *Energy, 77* (1), 859–870.

Qudrat-Ullah, H. (2015). Independent power (or pollution) producers? Electricity reforms and IPPs in Pakistan. *Energy, 83*(1), 240–251.

Qudrat-Ullah, H., & BaekSeo, S. (2010). How to do structural validity of a system dynamics type simulationmodel: The case of an energy policy model. *Energy Policy, 38*(5): 2216–224.

Qudrat-Ullah, H., BaekSeo S., & Brian, M, (2012). Improving high variable-low volume operations: An exploration into the lean product development. *International Journal of Technology Management, 57*(1/2/3), 49–69.

Qudrat-Ullah, H., & Davidsen, P. (2001). Understanding the dynamics of electricity supply, resources and pollution: Pakistan's case. *Energy, 26*(6), 595–606.

Ross, M. (2013). Environmental and economic consequences of Ontario's Green Energy Act 2013. Ontario Prosperity Initiative. Fraser Institute. http://www.fraserinstitute.org. Accessed on July 15, 2013.

Sterman, J. (2000). *Business dynamics: Systems thinking and modeling for a complex world.* New York: McGraw-Hill.

Stokes, C. (2013). The politics of renewable energy policies: The case of feed-in tariffs in Ontario, Canada. *Energy Policy, 56*, 490–500.

Chapter 7
Finale

> I never teach my pupils: I only attempt to provide conditions in
> which they can learn
>
> —Albert Einstein

Energy policy issues and challenges abound. The stock and flow perspective provides a unique and powerful set of tools to equip the policy decision makers better to take on these challenges. In fact, the system dynamics approach is well suited for the modeling and simulation of an integrated and holistic perspective on complex dynamic systems with the design of an effective and efficient energy policy that fully accounts for sociotechnical, economic, and environmental concerns. Let me conclude by delineating a roadmap for how, in a step-by-step approach, energy policy makers can improve their decision making by utilizing the stock and flow based modeling (i.e., system dynamics) approach.

7.1 Roadmap for Improved Decision Making in the Energy Policy Domain

At the outset of this book, we set the objective to aid energy policy makers to make improved decisions by embracing a stock and flow perspective. That is,

(i) Energy policy making is a complex dynamic problem (Chaps. 1 and 2).
(ii) The stock and flow perspective can effectively address energy policy issues by explaining the physics of underlying feedback loop structures of an energy system (Chaps. 3–5).
(iii) The stock and flow perspective has been successfully applied to solve various complex, dynamic issues of energy policy (Chap. 6).
(iv) For effective energy policy making, the embracing of the stock and flow perspective is the way forward (this chapter).

Please note that we are not claiming that reading this will make you an expert on system dynamics modeling or energy policy decision makers have to become expert on system dynamics modeling (great, if they are) but this brief book is to motivate

© The Author(s) 2016
H. Qudrat-Ullah, *The Physics of Stocks and Flows of Energy Systems*,
SpringerBriefs in Complexity, DOI 10.1007/978-3-319-24829-5_7

Fig. 7.1 Stock and flow
perspective and energy policy
decisions

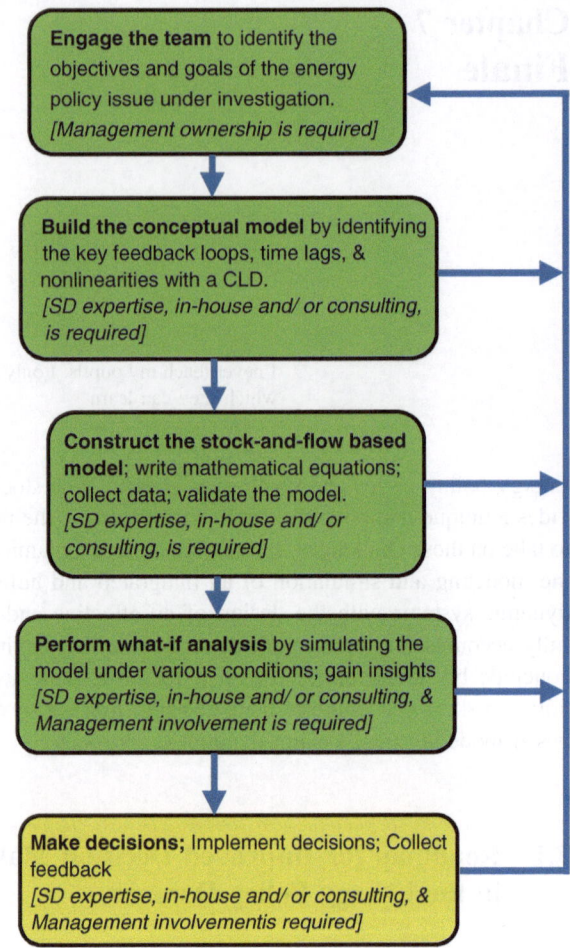

Engage the team to identify the objectives and goals of the energy policy issue under investigation.
[Management ownership is required]

Build the conceptual model by identifying the key feedback loops, time lags, & nonlinearities with a CLD.
[SD expertise, in-house and/ or consulting, is required]

Construct the stock-and-flow based model; write mathematical equations; collect data; validate the model.
[SD expertise, in-house and/ or consulting, is required]

Perform what-if analysis by simulating the model under various conditions; gain insights
[SD expertise, in-house and/ or consulting, & Management involvement is required]

Make decisions; Implement decisions; Collect feedback
[SD expertise, in-house and/ or consulting, & Management involvementis required]

you to take actions in your organization to instill the virtues of the stock and flow perspective. So, for dealing with complex, dynamic energy policy issues (i.e., evaluation, assessment, and design of energy policies and related issues), one can take these steps (Fig. 7.1), in an iterative manner.

7.2 A Few Personal Reflections

I was fortunate to have the opportunity to learn system dynamics methodology and embrace the stock and flow perspective about 15 years ago when I was a graduate student at the University of Bergen (Norway). Since then I have been developing and applying system dynamics models to a variety of complex dynamic problems including in the energy policy domain, with varying degrees of success (…yes, introduction of stock and flow perspective to bureaucratic mindset is not an easy task).

Energy policy challenges will continue to exist and so will the efforts of the research community to address them. In this context, a continuum of the innovation and growth in the development and use of models exist. Let me say it, "The utility of any model is in the eye of its users." No one is denying the usefulness of other modeling approaches (e.g., econometrics, optimization). In fact, my own experience shows that it is the combination of these methods and system dynamics approach that creates the best "solutions" for the complex, dynamic, and "wicked" problems and issues of the energy policy domain. You, the reader of this brief book, therefore, is encouraged to embrace the system dynamics approach, and then be the judge. I did it and have no regret at all ☺. Finally, I would appreciate your feedback on any concept, method, or material presented in this brief book, at: hassanqu@gmail.com.

Index